THE FIRE ENGINE

REMONT
5

THE
FIRE ENGINE
BY T.A. JACOBS

SMITHMARK

Water has long been the fire extinguishing agent. With greater knowledge of fire and its behavior, other substances have come into use as well: carbon dioxide, chemical foam and baking soda. While the means of discharging these substances range from buckets to hoses, to aerial drops from tanker planes, and pressurized discharge from fireboats, the chief means of extinguishing house, car and building fires is the fire truck.

The words 'fire engine' once referred to any machine that was used to put out fires. Contemporary parlance has narrowed that definition to mean the fire truck, as opposed to other apparatus for putting out flames. Even so, in studying its history, the progenitors of the modern fire engine must be included.

With its roots in the frantic hands of a primordial man hurling stones and debris at a campfire out of control, and its relatives, including the sprinkler system at the office, the box of baking soda on the shelf, and the giant tanker planes used to dowse huge fires in hard-to-reach places, the fire engine has become a sophisticated and versatile tool, indispensable to modern fire fighting.

Below: **Modern fire fighting has evolved far beyond horse-drawn siphonas. This Angloco Type B water tender, a first strike vehicle and the workhorse of Angloco's fire team, is equipped with water and foam extinguishing agents to handle domestic fires and hazardous materials.**

INTRODUCTION

Fire fighting is a battle for time—time to rescue the occupants of flaming buildings; time to prevent the spread of the fire; time to save what might be left of the victims' belongings. The proper equipment is essential, for there is no allowance for mistakes or second guessing.

Fire, both friend and enemy to mankind, has waged its own kind of warfare through the years. The Romans and Greeks of antiquity used fire to strike terror in and inflict mortal injury on their foes. Then again, more than 70 percent of Rome burned in the eight days of the Great Fire in 64 AD, and incalculable losses occurred when the library at Alexandria—packed with all the wisdom of the ancient world, on one-of-a-kind, irreplaceable tablets and scrolls—burned in the first century BC.

The study of fire has been systematized only in the last 100 years, but in that time, great strides have been made in fire safety and prevention, with advanced fire-retardant materials and building codes, as well as safety regulations, greatly reducing the likelihood of fire in modern homes.

Adequate fire-fighting equipment has been a critical need for humanity throughout history. The testimony of the 700 lives lost in the San Francisco earthquake and fire of 1906 remains grimly eloquent—as do the fire-charred Oakland Hills, just across the Bay from San Francisco, where property damage from the fire of 1991 exceeded the $190 million record set by the Great Chicago Fire of 1871.

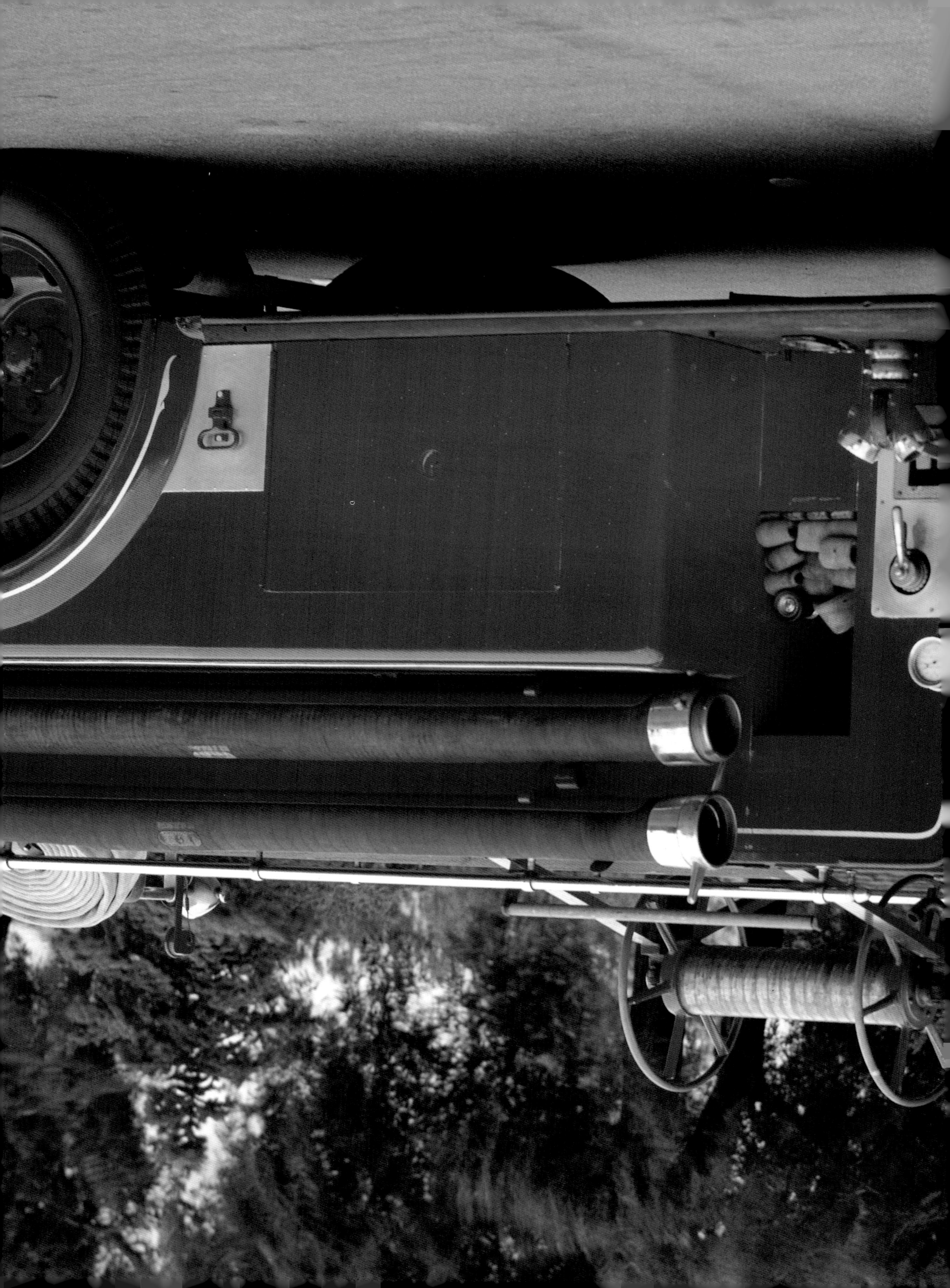

CONTENTS

This edition published in 1995 by SMITHMARK Publishers Inc., 16 East 32nd Street New York, New York 10016

SMITHMARK books are available for bulk purchase for sales promotion and premium use. For details write or telephone the Manager of Special Sales, SMITHMARK Publishers Inc., 16 East 32nd Street, New York, NY 10016. (212) 532-6600.

Produced by Brompton Books Corp.
15 Sherwood Place
Greenwich, CT 06830

ISBN 0-8317-3266-0

Printed in China

10 9 8 7 6 5 4 3

Designed by Tom Debolski
Edited by Randy Sean Schulman
Captioned by Lynne Piade

Page 1: **This KME Class A pumper rides on a Pemfab Imperial T-944A Series chassis and serves the Tullytown Fire Company of Bucks County, Pennsylvania.**

Page 2: **A Ford Cargo chassis is the basis of this Emergency One pumper.**

Gatefold: **This 1939 pumper is based on a Peterbilt S-100 chassis, the first chassis the company ever built. The body is by ON Hirst of Sacramento, California.**

The producers of this book wish to thank the following individuals and organizations without whom this book would not have been possible: American LaFrance Division of Figgie International; Michael Potter, Amertek Inc; Mick France, Angloco Inc; Sue Levy, CIGNA Museum and Art Collection; Tony Napolitano and Dave Gitchell, WS Darley; Marilyn Pasteur, Emergency One Inc; Alan Smith, Fire Fighting Vehicle Manufacturers Association; DA Haines, Ford Motor Company; John Shankland, Fulton & Wylie Ltd; Mr Kreuzer and Ms Olding, Iveco Magirus AG; Dick Schlicting and Amy Miller, Kenworth Truck Company; Tom Collier, Leyland DAF; Edward Raniszeski, Mack Trucks; MAN Nutzfahrzeuge AG, Dr Welsch, Mercedes-Benz AG; John McDonald, Navistar International; Keita Shima, Nissan Diesel Motor Co Ltd; Oshkosh Truck Company; Dick Gergel, Pemfab Trucks; Ken Elliott, Peterbilt Motors; Louis Helverson, The Free Library of Philadelphia; Barbara Weber, Pierce Manufacturing; Patricia Akre, San Francisco Public Library; Donald Wood, San Francisco State University; Katy Jane Mason, Simon-Dudley Ltd; Simon Gloster Saro Ltd; Nancy Immler, Sutphen Corporation; Scott Anderson, Volvo GM Heavy Truck Corporation.

Left: **When fighting major fires or neutralizing explosive situations, precisely-aimed extinguishing equipment is essential.**

Above: **Today's fire engines are measured against stringent safety guidelines. Enclosed cabs and bucket seats have improved conditions for fire fighters.**

Right: **Protective clothing has been developed for a variety of hazardous conditions.**

Below: **Aerials offer a superior advantage.**

FROM BUCKETS TO SQUIRT GUNS

Fire apparatus seem to develop sophistication most swiftly in cities. The reason is obvious once one considers the word 'conflagration,' which is defined by the *American Heritage Dictionary* as a 'large and destructive fire.'

In terms of harsh reality, a conflagration generally means any fire that involves a large area and causes massive damage. Cities have always been the most susceptible venues for these kinds of disasters due to their high concentration of buildings and gas lines. City conflagrations have the natural potential to threaten large numbers of human lives.

Add into that the fact that cities have always been the prime targets in warfare: witness the destruction of Jerusalem at the hands of the Romans, and more recently, the terrors of World War II, in which 65 Japanese cities were destroyed an average of 50 percent, while 54 of Germany's largest cities received an average of 40 percent destruction—Dresden representing the largest fire *per se* of that war, with varying estimates of human deaths in that firestorm ranging from 35,000 to 135,000.

The earliest man-made records of fire-fighting equipment tend to come from urban locales. We do not know who, or what was used in extinguishing fires 7000 years ago—they probably threw dirt on fires just as the average

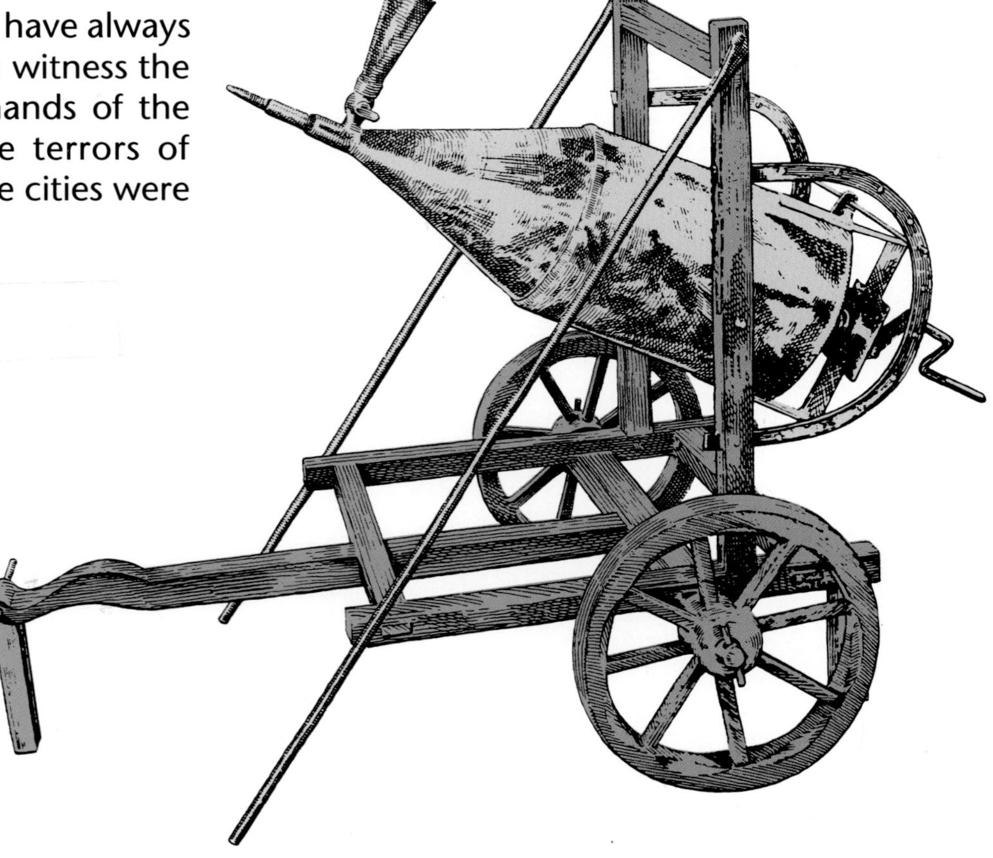

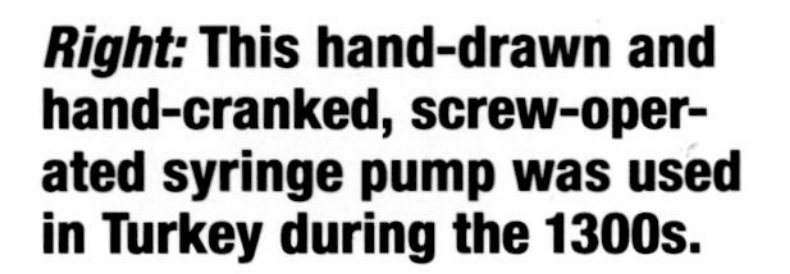

Right: **This hand-drawn and hand-cranked, screw-operated syringe pump was used in Turkey during the 1300s.**

Opposite: **Built in 1856 by Hunneman & Company, this lovingly restored hand pumper is one of the 750 built by the Massachusetts-based company.**

camper does today, and somewhere along the line, someone discovered that water was even more effective.

Once that notion caught on, it was just a short leap of logic to carrying the water used to dowse the fire in an urn, bowl or bucket.

In the fourth century BC, an Alexandrian Greek named Ctesibius invented a pump with inlet valves at the bottom of each of its two cylinders. The pump was made to be set in a stream or a tub of water, and was operated by rocking an arm that caused the pistons to reciprocate in an up-and-down motion, forcing a jet of water through a central opening in the top of the pump.

Pliny (62-113 BC) makes mention of the use of fire engines in the Rome of his time, and Hero of Alexandria (about 150 BC) described a fire engine with two pistons worked by means of a reciprocating lever.

These early syphon-pumps became known as 'siphonas,' and would represent state-of-the-art fire-fighting apparatus for several centuries. Improvements were made over the years, most notably the addition of a tub or reservoir of water on the pump carriage. This reservoir was refilled most often by means of a bucket brigade.

These siphona-style pumps were used in Germany as late as the sixteenth century, and ranged in size and configuration from one-cylinder washtub-size pumps—man-hauled on runners or cart wheels—to a behemoth in Nuremburg that was pulled by two horses and had to be operated by 28 men.

By 1400 AD, Turkey was using syringe pumps, which were hand-cranked and worked by means of an internal screw mechanism. They were often mounted on two-wheeled carts that were drawn by hand.

At the Great Fire of London in September of 1666, the fire equipment included siphonas and syringe pumps. The syringes were three to four feet long, and two and a half to three inches in diameter.

The syringe's effectiveness was on par with that of the siphona. Both helped, but often enough became victims of the flames they fought, as they were too weak for anything but close-in operation.

While London had burned before—in 798, 982 and 1212—the Great Fire of 1666 emblazoned the question of better fire-fighting apparatus upon the European consciousness. Perhaps it was an early spark of the mechanical inventiveness that would, over two centuries, conflagrate into the roaring furnaces and boilers of the Industrial Revolution.

Such fires as had occurred in Dresden, which was completely consumed in 1491; Leith, leveled in 1544; Cork, almost completely burned in 1612 and again in 1622; and Stratford-Upon-Avon, burned in 1614, added a sense of urgency to the quest for better fire equipment.

In 1673, an answer came from Holland. Jan and Nicholas Van Der Heiden, father and son, created a fairly reliable flexible hose by sewing strips of leather together longitudinally, using strong linen thread to close the seams. They called their new creation 'water snakes' (the English would later dub the invention 'hose,' after the popular word for stocking).

The Van Der Heidens' hose leaked notoriously, but it was more useful than the bucket brigade, and delivered most of the water to the fire. Father and son applied their hoses to their own relatively powerful siphona-style pumps, using them for discharge and for suction. This meant that, for the first time, the pump would not have to be placed so close to the fire, and water could be drawn from a distant source without the aid of a bucket brigade.

Larger pumps were being developed, too—requiring the services of 40 or 50 men to operate them. Even so, the word was slow getting out to the rest of the world.

Houses of Kindling

Houses and public buildings in the North American Colonies were mostly of wood-frame construction, and the bucket brigade prevailed as the chief Colonial fire-fighting method into the seventeenth century. Boston had a terrible fire in the 1650s—the same urgency that would

***Right: Old Brass Backs* got its name from the builder's lavish use of brass on the box of the engine. It was the first fire engine built in New York and Thomas Lote delivered it to the city in 1743.**

Left: **This was New York's first fire engine, manufactured by Richard Newsham of London in 1731.**

Right: **Engine Company No 6 waited 77 years for a fire engine and in 1842 they received *Big Six* from John Agnew of Philadelphia. This engine is an example of the 'Philadelphia' style of apparatus design.**

be borne in the smoldering rubble of London was borne amidst the ashes of Boston.

In 1654, Bostonians commissioned Joseph Jencks, an ironsmith from Lynn, Massachusetts, to make an 'engine' for fighting fires. Authorities say that it was probably a syringe pump, supplied with water by a bucket brigade.

Apparently, Jencks' engine was not up to snuff, for the City of Boston ordered a water pumper from London in 1676. Before that pumper arrived, Boston suffered another major fire, losing 50 houses, several warehouses and a church, and would have lost more but for a sudden, torrential rain.

The pump from London arrived in 1678. It was known as a 'fire tub,' and consisted of a box three feet long and 18 inches wide, and had carrying handles and a nozzle out in front. A one-cylinder affair, it served the city for 30 years. In fact, Boston ordered more tubs, and by 1715, there were six fire tubs in Boston.

The rest of the Colonies still relied on bucket brigades and prayer, prayer being by far the more effective of the two.

Better Pumps and Ways to Pump Them

In 1720, an Englishman named Richard Newsham upgraded existing ideas on pump making and produced a man-powered machine that delivered 70 to 170 gallons per minute, or gpm (144 to 265 liters per minute, or lpm). Newsham added to the appeal of his apparatus by putting wheels on it.

The story goes that Newsham might have become just another inventor but for the intervention of King George I, who felt that Newsham's engine was just the thing—for the Royal Gardens!

Be that as it may, Newsham's products became all the rage. Newsham's engines were not all of one type, however. In 1731, the City of Philadelphia ordered two Newshams from London. One of the machines had a two-piston pump with side-stroke levers for 'galley' pumping, and foot treadles for operation in confined quarters. The other Newsham had a rotary pump with cranks on either side.

Newshams were very popular, but Yankee ingenuity even then reared its head, and Thomas Lote, of New York, invented what has been called the first practical American fire engine. It was a knock-off of Newsham's work, but hearkened back to the 'siphona' of old. *Old Brass Backs*, as it was known, had wheels and a fixed nozzle, but improved on the Newsham by having its pump-handles situated front and rear, which was more suited to city use. The front-and-rear pump handles came to dominate American designs for many years.

Indeed, Richard Mason of Philadelphia produced an engine that rivalled Newsham's work in 1768. Built for the Northern Liberty Fire Company of Philadelphia, this engine had end-stroke handles, and established Mason's career, as he went on to become the most respected fire engine producer of late eighteenth century America.

Hand-operated engines grew in sophistication. Philadelphia's Pat Lyon developed powerful, reliable, double-piston end-stroke engines that became famous for their efficient reliability. These engines became known as 'Philadelphia' style engines. At first, they were fed by a bucket brigade, but were upgraded to hose-suction feed in 1810. From 1835 to 1860, John Agnew, another Philadelphian, continued to develop the Philadelphia style machines, increasing the size and maneuverability of the engines until they became the very best that hand-pumped machines could be.

Right: **This 1854 lithograph depicts several models of vintage fire apparatus i[illegible] eration, *from left to rig[illegible]* gooseneck or 'Old New [illegible] style engine; a 'piano box' style engine; a 'Philadelphia' style engine; and a four-wheel hose carriage.**

Left below: **This 'piano box' engine was built in 1853 and was one of the best geared and fastest engines of its day.**

Below: **A close-up view of one of the two pistons on a vintage end-stroke engine.**

In the United States alone, several different styles began to emerge. There was the 'New England' style, produced by Hunneman & Company of West Roxbury, Massachusetts—a very plain machine with a powerful pump and an innovative 'goose neck' frame that allowed for weight savings and good maneuverability. There was also the 'New York' style, which followed Newsham's designs closely, with a rear condenser case and side-stroke pump handles. The New York style allowed for carrying several lengths of extra hose as well, but never achieved the pumping power of its contemporaries.

New York then came up with several variations of the Philadelphia-style engine: these included the 'Shanghai,' or 'pagoda' style, engine developed by James Smith and William Torboss.

By 1850, the New York style engine was replaced outright by the 'piano box,' developed and perfected by L Button, of Waterford, New York, in the mid-1830s. This powerful engine, with its heavily framed, upright side-stroke handles, came to define the notion of a hand pumper in New York.

Another piano box style was Henry Waterman's 'hay wagon,' developed in 1842. This pumper's huge pistons took so much exertion to pump that it was soon dubbed 'the man killer,' and production soon halted. A good 'piano box' could throw water 150 feet (45 meters) into the air.

There were, of course, rotary-pump engines—one type was the side-crank or 'coffee mill' style, and the other was the windlass, or 'cider press' style. Neither rotary configuration gained much popularity until steam technology came into the picture, and could take full advantage of the rotary pump's dynamics.

For some, steam was a long time coming. As good as hand pumpers became, too much damage was incurred because of inadequate equipment.

Right: **This restored Hunneman hand pumper is an example of the 'New England' style of end-stroke fire engines produced in the nineteenth century.**

ARTMENT
S.R. SPINNEY
N.14

THE COMING OF STEAM AND PROFESSIONALS

The first steam fire engine was built in England, around 1829, by John Braithwaite. It was an offspring of existing steam-driven pumps that were used for pumping water out of mines. The pump used steam to push a piston, which in turn pushed water through an outlet valve. A bell-like surge chamber allowed for a smooth flow of water.

The general idea became the framework for a revolution in fire-fighting equipment—including the surge chamber, which was a highly-polished brass or chrome dome. It was as much a decorative item as a necessity. The dome languished for 25 years before steam fire engines began to take their rightful place at the vanguard of fire equipment.

This was largely because the volunteer companies were steady customers for fire equipment, and they demanded hand-pumpers—both for their own vanity and for their own survival.

In Europe and in North America, volunteers made it almost impossible for early inventors to find backers for their steam-engine projects. John Ericsson, who would go on to design the Union North's famous ironclad battleship *The Monitor* in the Civil War, won a New York competition for fire engine design in 1841.

Ericsson's steamer was crude, but weighed just two and one-half tons (2500 kilograms)—a ballerina compared to subsequent iron monsters, and was easily pulled and highly maneuverable.

The first steam fire engine actually used in the US was *The Exterminator*, an eight-ton (8000-kilogram) monster built by Paul Hodge in 1841. It was manned by the Pearl Hose Company No 28 of New York, but produced such a low volume of water for the work it took to move it to the site of a fire that *The Exterminator* was eliminated after only a few months in service.

Other attempts were made with little success. Finally, during London's wharf fire of 1861, civic officials had had enough of loose organization and volunteer antics. Similarly, in 1853, the mayor and city council of Cincinnati, Ohio were perturbed at riots and inefficiencies involving firemen. Both instances concluded in resolve: replace the current fire-fighting system.

Such companies as Carmichael were regularly manufacturing state-of-the-art fire engines in the UK, but the situation was slightly different in North America.

While London was to establish governmental precedents implementing the organization of better-staffed, more professional and better-equipped fire companies, Cincinnati took the

***Above: The Pioneer*, built in 1857, was Reaney & Neafie's first steam fire engine. Today it is part of the CIGNA Museum and Art Collection.**

direct approach: they commissioned the design and manufacture of a steam fire engine. Abel Shaw and Alexander Latta were chosen for the work. Their handiwork was dubbed the *Uncle Joe Ross*, weighing 22,000 pounds (9980 kg). It was propelled by steam power and boosted by four horses ridden artillery style. It had good pressure, and could throw its stream of water 225 feet (69 meters).

The volunteers were ripe for a pumping contest—such events having been their idea of a party. This was deadly serious, however—the impetus behind the *Uncle Joe Ross* was specifically to reduce the number of men in fire departments: the fewer fire fighters, the more controllable they would be.

While the volunteers, using a 'man killer' pump, managed to get their stream higher and further than the steamer, they collapsed in exhaustion while the steamer continued to pump an adequate stream of water, and could apparently do so for *hours*.

The citizens of Cincinnati were so impressed with the *Uncle Joe Ross* that they wanted more steamers and contributed to a fund to purchase a second steamer, also a Latta design, called *The Citizens' Gift*.

Cincinnati organized the first professional fire department in the US on 1 April 1853. The

age of volunteerism was drawing to a close, and the age of professional companies, equipped with the latest fire engines, was dawning.

Thus, in the space of one decade, cracks in the volunteer establishment were opened on both sides of the Atlantic, and while volunteerism stubbornly resisted steam engines for the following 50 years, it was clearly a losing battle, with steamers proliferating right up to, and *into*, the age of internal combustion engines.

The Industrial Revolution was a contributing factor. As people throughout the world were lured from the countryside to the cities by the false promises of high pay in the sooty, foul, dark factories of the period, the world was becoming rapidly urbanized.

Volunteers manning hand-pumpers couldn't generate enough pressure to dowse flames that were frequently located above the third story level. A strong, steady stream, and fire fighters who would concentrate on the job, rather than rivalry, were needed. The steamer and the professional fire department provided both.

Latta engines were in service in St Louis by 1855, and had been demonstrated in several Eastern cities. Some volunteer companies ordered steamers (hand-drawn, naturally, to keep their manpower up), and many professional companies acquired them as well.

Types of Equipment

Probably the first auxiliary piece of equipment used by fire companies was the water bucket. Having been first the main 'engine,' with the advent of water pumps, the water bucket became the main source of feeding the pump with water.

The bucket was replaced with the water main system and the hose. The hose was developed first, but the water main system was the means by which hoses could be increased in effectiveness. One early water main system was built in Boston in 1796. This system was made of wood, and was gradually upgraded to an iron main system with pumping stations.

This made the periodic location of fire hydrants possible. With the water main system came the need for hoses that could handle higher water pressures. The Van Der Heiden leather hose would not do.

Left: The gleaming brass on this lovingly maintained steamer recalls an era long past.

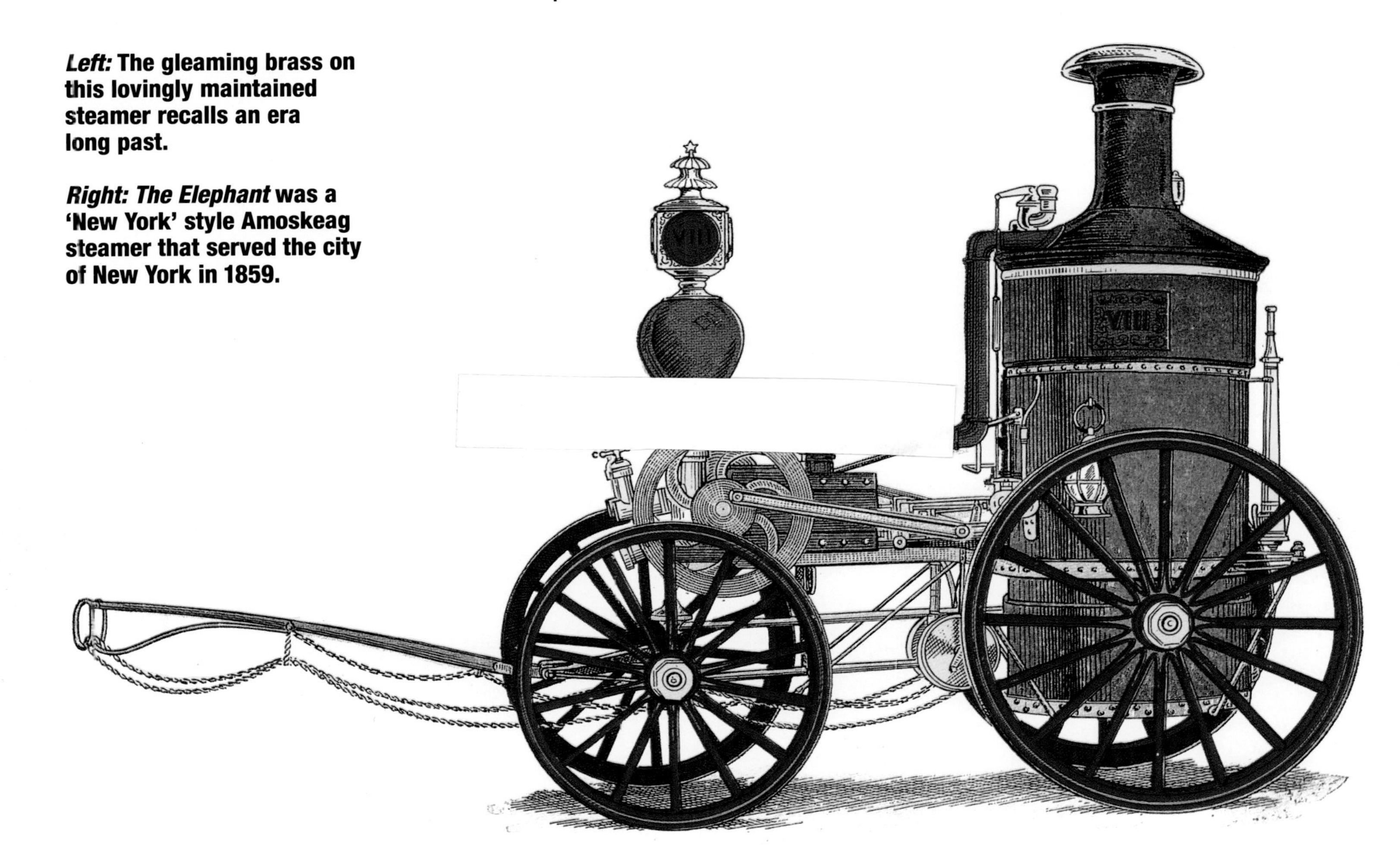

Right: The Elephant was a 'New York' style Amoskeag steamer that served the city of New York in 1859.

11956

In 1808, the Philadelphia firm of Sellers and Pennock developed leather hose that replaced linen seaming thread with iron (later, copper) rivets, for a tight, leak-proof seam. This was the standard until 1870, when rubber-lined cotton hose took over and remained the standard for fire hoses until recently, when advances in synthetics and various combination substances have made for yet another standard.

Hoses of course have necessitated a means of carrying them—the hose truck. Previous to the nineteenth century, most firemen carried their hoses on their shoulders, with a few exceptions. Then, simple buckboard-type wagons were used. Around the turn of the nineteenth century, the notion of equipping fire engines with hose reels came into vogue, and then, because such reels were too small to be of much use, two-wheeled carts became the norm for carrying hose. These carts were called 'turtlebacks' because of their broad curvature in profile. The hose was laid on a series of flat windings.

Even so, the two-wheeled turtleback cart didn't have the hose capacity of the old wagons. In approximately 1810, the 'jumper' hose cart was developed. This vehicle was composed of a mammoth hose reel and two wheels. It could easily be jumped over curbs (hence its name), and had more hose capacity than the turtleback.

About 1812, the four-wheel version of the jumper was developed. Properly called a 'hose carriage,' it was lightly built, with a single arching spring supporting either side, and the requisite huge hose reel.

When professional fire departments began to spring up in the 1850s, horses saw increasing service pulling the apparatus that had been hauled by volunteers.

In the 1870s, the idea of the hose wagon was reintroduced. The new cotton and rubber hoses suffered from mildew when wound on a reel, so a more capacious vehicle was needed. The wagons also allowed for carrying firemen, who could thus arrive at a fire without the ex-

***Left:* Collectors of antique fire apparatus have reconditioned and preserved beauties like this for all time.**

haustion of having run much of the way. In snowy climes, these wagons were built with sled runners: in New England such hose sleds were called 'pungs.'

The 1870s also saw development of what were called 'chemical wagons.' These carried tanks of soda and acid, positioned either vertically or horizontally (the horizonal tanks were easier to charge), and were used to chase sparks and embers or fight small beginning fires. When combined, the soda and acid made a frothy, wet mixture that smothered fires easily.

By 1890, chemical and hose wagons were combined to perform a double function, and with the advent of internal combustion-powered trucks, were replaced by a triple function vehicle, a combination pump, hose and chemical (or booster tank) truck.

Taking It to the Flames

Even with better equipment, fire fighters had to access the fires they were to fight. In rural areas of the time there were no effective means to cut off a grass fire or a forest fire. Men and horses often lost the race to the next fire gap, and buckets and thin streams of water were all but useless against such infernos.

In cities of the nineteenth century, a desperate struggle was also under way: growth was vertical, and so were fires; that meant ladders had to be used by fire fighters, and many times, the ladders lost the battle.

The first hook and ladder apparatus to be used in the US was bought by Philadelphia Fire Company Number One in 1799. The ladders were for accessing higher-story fires, and the hooks were for pulling down partition walls to prevent the spread of fire.

Such early 'hook and ladders' were simple affairs, consisting of long wagons loaded with ladders and poles or chains equipped with hooks. They also had dozens of water buckets hanging from their undersides.

Early on, a legendary 'nameless genius' added a turnable set of wheels to the rear of the wagon to help it turn corners. These were mostly drawn by hand, until professional companies demanded horse-drawn versions in the 1850s. Soon after, a seat in front, for the driver, and a seat and steering wheel in back—for the tillerman who steered the rear wheels—were added.

Left: **This style of hose cart became known as the 'jumper' by virtue of the ease with which it could be jumped over curbs.**

Right: **An 1891 Silsby steam fire engine now part of the CIGNA Museum and Art Collection.**

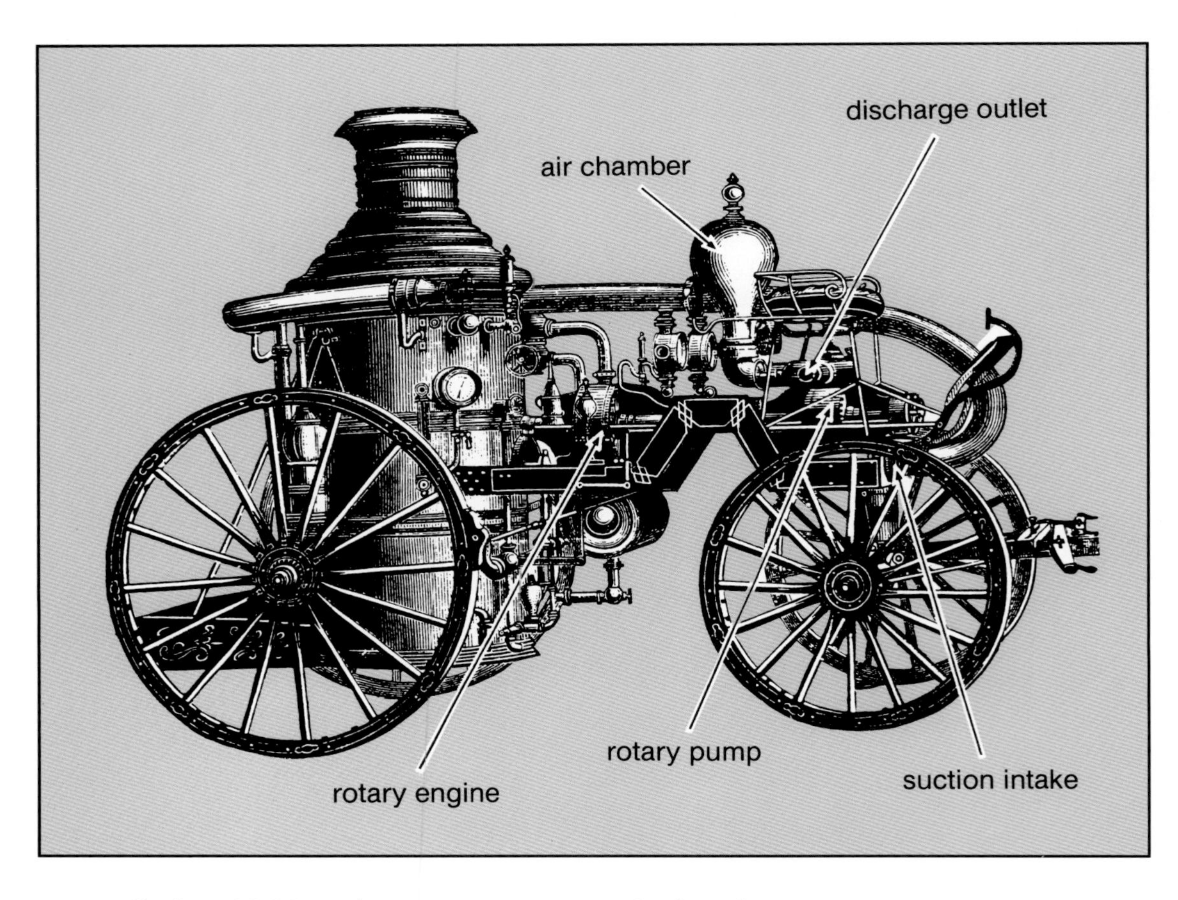

Left: **The heart of this Silsby steamer is a horizontal rotary pump.**

Below right: **This handsome steamer is trimmed in brass and copper.**

Until the 1880s, these wagons carried only 'ground ladders' of up to 75 feet in length. Their butt ends all too often slipped, dropping a hose-laden fireman to his death. Early attempts at more stable ladder arrangements resulted in inventions such as the Skinner, which used a series of pulleys to raise the ladder, hinging upward from the wagon bed; and the Scott-Uda, which operated like a scissors jack, and which also proved dangerously unreliable.

Daniel Hayes of the San Francisco Fire Department patented a safe aerial ladder in 1868, but, like the steam-pumped fire engine, this languished for over a decade. Only when the La- France Company began manufacture of the 'Hayes Aerial' in 1881 did firemen over a wide area have access to a safe aerial ladder.

The Hayes ladder was attached to a turntable on the bed of the truck. The ladder was raised by a highly controllable worm-gear, and its extension was controlled by a well-engineered crank system.

This system was copied by many firms, and was eventually improved by the Dahill air hoist and spring hoist of the early twentieth century. These of course gave way to hydraulic and 'power takeoff' hoists, which in turn saw the advent of elevating, aerial platforms and Snorkel trucks—which were introduced to the fire fighting world by the Chicago Fire Department in 1958.

The Glories of Steam

With the success of the Latta steamers various companies began manufacturing steam fire engines. The LaFrance Company manufactured both rotary pump and piston pump models. Amoskeag built mainly piston pump models. Then there were Ahrens, Clapp & Jones, Reaney & Neafie and others.

Reanie & Neafie produced 40 engines by 1869. Their first, built in 1857, weighed about three tons (2724 kg). Its vertical fire tube boiler fed steam to its single pump and cylinder. It could raise full steam in eight minutes and was utilized for 40 years.

The average steamer at the end of the nineteenth century could pump about 500 gallons per minute (gpm) (1892 lpm) of water, but some could produce 1000 gmp (3785 lpm). The now-ubiquitous fire dog, usually a Dalmatian, has its roots in this period.

The dogs were station mascots, usually, and were chosen for their ability to chase the fire engine to the fire, and to enliven the comparatively boring times between fires. When horses were no longer used, but gasoline and diesel engines had taken over, the dogs were retained, initially because firemen missed their horses and sought to retain some of the warmth of the earlier days. Now the dogs are a tradition.

Several basic types of steam fire engines were a result of the small multitude of engine makers extant in the mid-to-late nineteenth century.

There were those with a horizontal piston and pump; a single vertical piston and pump; a double vertical piston and pump; a vertical piston and standard horizontal pump; a rotary pump and single vertical piston, a configuration called a 'mongrel'; a rotary pump with a single horizontal pump (another mongrel); and a horizontal rotary pump.

The biggest challenge to steam fire engine builders was making a machine that was light and maneuverable, while still being large enough to surpass the best hand pumps. In the early days of apparatus design it was a tradeoff of weight for water pressure and vice versa.

The introduction of high-tensile steel made it possible to produce powerful but lightweight engines, and better engineering made for better and more reliable pumps. There was a lot of bleed-through from the locomotive and marine steam engine industries.

Such developments took money and experience to acquire and apply. By 1860, most of the small manufacturers had dropped out, leaving the field to the larger companies, like Ahrens, LaFrance and Amoskeag.

Some smaller manufacturers did persist, however. Among one of the more interesting

was Silsby Manufacturing Company of Seneca Falls, New York. The company produced their first steamer in 1856, and achieved the peak of their form in the 1870s.

Silsby specialized in steam-driven rotary-pump engines. The advantages of the rotary pump were that they had unusual stability of operation, imparted a completely even flow of water and had few moving parts to wear down or break.

Clapp and Jones, of Hudson, New York, designed a better boiler in the 1870s-1880s, based on the water-tube principle, unlike the 'gunpowder,' or fire-tube boilers previously used on steamers. The water tube boiler was far more reliable and also conserved steam for re-use. This boiler design became a standard.

The Amoskeag Locomotive Works of New Hampshire produced 853 steamers between 1859 and 1913. They produced few hand-drawn engines, many horse-drawn engines and some self-propelled engines.

Amoskeag invented the differential gear, variations of which are still used in automobiles

and trucks. This makes it possible for powered vehicles to turn corners without spinning one of the rear wheels. The differential gear allows a different rate of rotation under power to occur between the inside and outside wheel in turns.

By the end of the nineteenth century, many manufacturers were conglomerating. In 1891, Silsby, Ahrens, Clapp & Jones and Button formed the American Fire Engine Company. The new firm produced the Metropolitan Steamer, a vertical piston design, and one of the last of the famous steamer designs.

It featured a 'gooseneck' carriage, inspired by the conveyances of seventeenth-century French royalty, whose ornate carriages needed the sort of structural strength that the heavy steam fire engines did. Even today, heavy trailers feature this gooseneck design.

American combined with LaFrance, Rumsey, Gleason and others in 1900, to form the International Fire Engine Company. This name was changed to American La France in 1904 to cash in on the name value of the two most prominent names in the combination.

Left: **The dawn of the twentieth century saw the beginning of the end for steam fire engines. Gasoline-fueled internal combustion engines, such as this one manufactured by the Waterous Engine Works in 1909, replaced the old steamers.**

NEW WAYS, NEW EQUIPMENT

The era of dominance for volunteers was largely gone by the turn of the twentieth century. Even so, in various towns and cities around the world, especially in England, there still remained vestiges of bygone days. While steamers were now part of common fire-fighting vocabulary, many fire-fighting squads used them only for the direst emergencies, relying on high-pressure hydrant water where available.

In such cases, the company took only their hose wagon. Other companies took manual pumpers, most often as transportation for the men; still others took the manuals and made a point of using them, high-pressure hydrant water notwithstanding.

Organization of fire departments to make use of fire engines was key. Two success stories of the early twentieth century come from Germany.

The Cologne professional fire brigade was 153 strong as of 1906, with fire officers, a telegrapher and 16 foremen. Its 26 horses included two assigned to ambulance duty. The crew were divided among three large stations and one small station, with men specifically assigned as drivers, which avoided confusion when the alarms sounded.

Cologne was outstanding in that it also had a regularly maintained auxiliary comprising three officers, five foremen, 51 men and two horses. This auxiliary was housed in fire department tenements to be always available as an immediate reserve force.

There was also a 295-member suburban force of volunteers, equipped by the municipal department, and overseen from within their well-organized ranks by five officers per each 50 firemen.

Nuremberg was considered the most economically organized and efficient fire service in

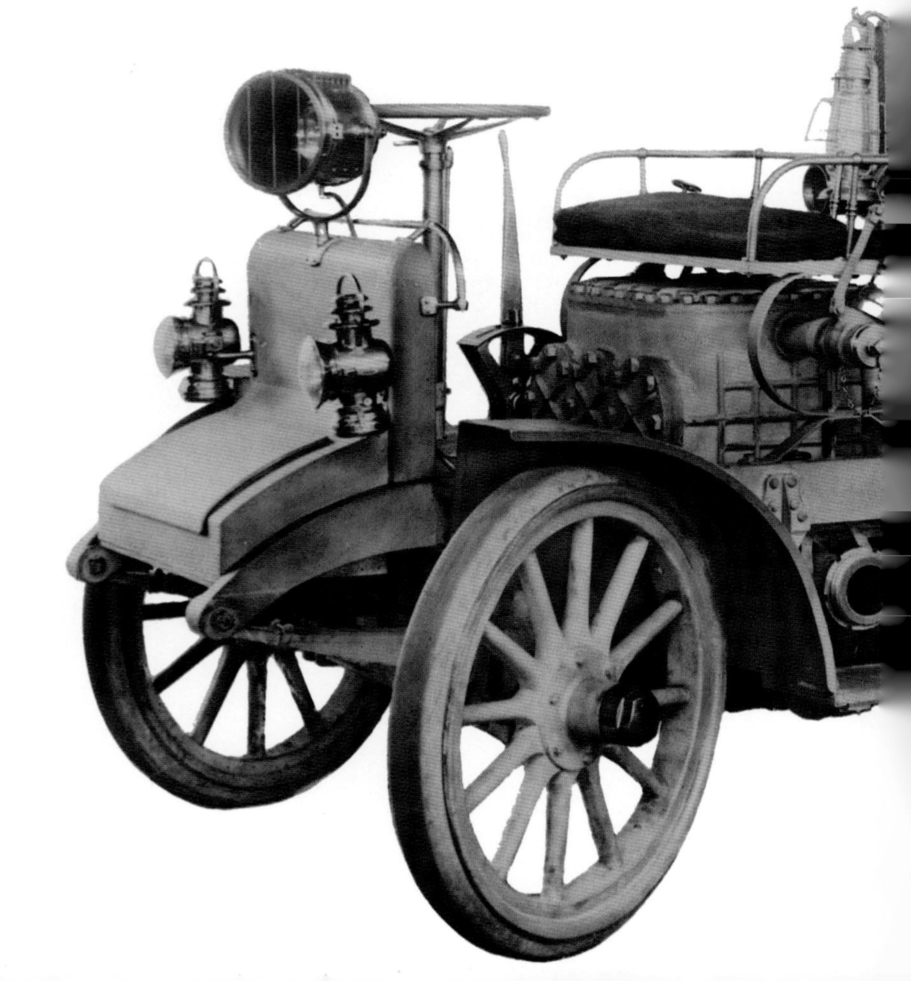

Above: **Lightweight, hand-drawn, gasoline-powered engines, such as the one shown here, replaced heavy steamers in many fire departments at the turn of the century.**

Left: **The Lutweiler Pump and Engineering Company of Rochester, New York built this pumper in 1909. It is one of the earliest rigs to use the same power source for propulsion and for pumping.**

FIRE DEP

central Europe in the first decade of the twentieth century. In 1907, for example, this city of approximately 261,000 inhabitants (at the time) got the following brigade service for £6300: 156 highly-trained firemen in a retained force, and two well-trained volunteer brigades of 354 men in all; an auxiliary of 18 suburban brigades totaling 1080 men; five factory brigades totaling 71 men; and a dozen clerks.

The retained brigade was comprised of city employees who worked in city shops that were purposely located next to the respective stations—thus the city got two professionals for the price of each man's wages.

The Nuremberg fire brigades were well-equipped with ladder trucks, hose trucks, chemical engines and steamers. The whole setup had been in place for 50 years as of the turn of the century, and was highly successful in its fire protection role.

Equipment and techniques improved by the relative dearth of major fires in the British Isles and Europe from the mid-nineteenth to the early twentieth century.

Even so, Australia suffered a terrible loss in the Flinders Lane fire of 1897, and Toronto, Boston, Hoboken, Atlanta, Baltimore, New York and San Francisco all suffered disastrous fires in that same time frame.

Though the US at the time had a fire-fighting force four times that of Germany or France and three times that of England, the American tendency to build mostly of wood, even in cities, and climatic extremes, combined with overcrowding in immigrant neighborhoods, all contributed to 'tinderbox' conditions.

These pages: **This immaculately restored gasoline-powered engine, *The JW Pennington*, was originally built by the Waterous Engine Works Company of St Paul, Minnesota in 1906. Note the gleaming brass crank handle *(left)* used to start the pump motor.**

Internal Combustion Comes of Age

With the success of the internal combustion engine and the advent of the automobile, the role of steam in powering water pumps came into question—indeed, the horse's role as the prime mover of fire apparatus came into question, too.

The internal combustion engine was relatively light and compact, and yet powerful. Likewise, too, it had its early rivals—the electric motor and automotive steam powerplants.

In such locales as the US, steam and electric automotive power became obsolete by the 1920s, because neither was as easily maintained or capable of as wide a range of performance as the internal combustion engine. In the British Isles, however, some steam designs were used well into the 1950s because of the abundance (and hence, low cost) of coal in the United Kingdom.

Even so, internal combustion had carried the day in automotive and truck transport by 1920. With such manufacturers as AEC, Leyland, Albion, Foden and Dennis in the British Isles; Mack, White, FWD, General Motors, Ford and others in the US; and Daimler, Benz, Saurer, Fiat and others on the European continent, there was an abundance of truck manufacturers who made self-propelled vehicles in the first two decades of the twentieth century.

Many of these truck manufacturers were more than willing to collaborate with such fire equipment manufacturers as American LaFrance in supplying what promised to be a lucrative offshoot of the auto and truck market.

Above: **The Ford Model T chassis provided the motive power for many early self-propelled fire apparatus.**

Right: **A detail of a mid-1920s American LaFrance pumper.**

Above: **An American LaFrance pumper from the 1920s, used by Fairfield, California.**

Right: **Mill Valley, California still uses this circa-1920 American LaFrance pumper as a parade piece. It's shown at a Fourth of July parade in neighboring Larkspur.**

Left: **A 1921 Packard with a Pirsch pump, photographed at an auction in Sparta, Wisconsin in 1988.**

Some of the first attempts at linking the new self-propulsion systems with fire fighting resulted in such hybrids as the tractor-steamer combinations of 1910 to 1920.

At the same time, the concept of fire research was gaining ground on an international front. The United States established a good footing in this field when engineer William Merrill was hired by a consortium of fire insurance companies to inspect the electrical installation at the World's Columbian Exposition in Chicago, in 1893.

As a result of this work, Merrill became intrigued with the fire hazards posed by electrical systems, and persuaded his employers to sponsor laboratory tests of various electrical devices, and establish an approval system to guide the insurance companies in writing fire insurance. Underwriters' Laboratories, the world's largest safety testing organization, was the result of Merrill's efforts.

In Great Britain, London's Cripplegate fire of 1897 gave impetus to the creation of the British Fire Prevention Committee, which tested fire apparatus and fire-fighting methods.

From those beginnings, interest in fire research established such late twentieth-century organizations as the Joint Fire Research Committee in the UK, the National Fire Protection Association in the US, and organizations set up by Underwriters' Laboratories of Canada, the Canadian Standards Association and the Canadian National Research Council.

Devastating fires throughout the world only hastened the urgency. Melbourne, Australia suffered enormous fire losses in the notorious

Left, top and bottom: **The restoration of this 1920s Seagrave truck is the handiwork of Terry and Joel Meyers of the Twin Hills Fire Department in Sebastopol, California, who restored the vehicle in 1989.**

Right: **When this siren sounded, people knew that help was on the way.**

Below: **Albany, California operated this circa-1940 Seagrave pumper.**

Flinders Lane conflagration of 1895, and in the destruction of a large knitting mill in 1925. Halifax, Nova Scotia, suffered 1500 deaths when a munitions ship exploded in the harbor, levelling 75 acres of the city. Baltimore suffered a $150 million loss in 1904, and the Great Earthquake in San Francisco left fire in its wake causing $400 million in damage and 400 lives lost.

Not all devastating fires were large. Some of the smallest fires caused inordinate numbers of casualties: the Iroquois Theater Fire of Chicago in 1903 killed 602 and led to more stringent regulations for theaters worldwide. The terrible Triangle Shirt Waist Factory Fire of New York, in 1911, was a poorly safeguarded 'sweat shop' that burst into flames. Because one exit was too small and the other was locked, many people fled the fire by jumping through windows to their deaths. The death toll was 145.

Not all fires were confined to land. In 1904, the excursion steamboat *General Slocum* burned just off the shoreline of New York City, with a loss of 1030 lives.

Such horrors convinced fire officials worldwide that better standards were necessary, and one of those standards was speed in getting to the fire. Another was power to put it out.

The steam fire engine during the turn of the century spewed a respectable stream of water—1250 gpm (4732 lpm), on par with a present-day medium-sized engine. The problem was that, even with the best of them, steamers were still pretty cumbersome for their fire-fighting ability. Horses could haul only so much—if you wanted to haul more equipment, you needed more horses, and the prospect of a

These pages and overleaf: **This 1941 Darley pumper is built on a Ford chassis that was specially modified for operation on narrow mountain roads.**

9·1·1
EMERGENCY

COLMA FIRE DIST.
CASTRO VALLEY

***Left above:* This White fire truck saw service in the 1940s.**

***Above:* An American LaFrance pumper from about 1940. Note the enclosed overhead ladder rack.**

***Left:* A Chevrolet chassis is the foundation of this sleek pumper.**

small cavalry of horses charging through city streets brought to mind the reality of the amount of time necessary to harness a herd of horses.

Quicker response was the name of the game. The internal combustion engine would not prove itself in this regard, however, until the 1920s, when self-starters became more generally available. Oftentimes, while the truck could outrace the horse, it had trouble getting started.

Firemen were still trying to crank-start their engines while other firemen were already racing to the fire, hauled by horses. That didn't stop the development of the fire engine as we know it today. A few automakers built fire engines, but most merely supplied the chassis for fire engines.

Fire engines are often made to order: they also require heavier powerplants than most autos and trucks, as they have to power both the running gear and the fire-fighting apparatus. Companies such as American LaFrance and Wa-

terous Engine Works were among the first to produce internal-combustion-powered fire engines in the US.

American LaFrance built a combination hose and chemical truck for Niagara Engine Company of New London, Connecticut in October, 1903; Waterous built a two-engine pumper (one hauled the truck, the other did the pumping) for the Radnor Fire Company of Wayne, Pennsylvania in 1906. One Waterous motorized engine of 1907 had a capacity of 600 gallons (2271 liters) per minute.

On the other hand, the St Louis Fire Department was using a battery-powered electric combination truck as early as 1897.

Such advances also begot rivalries. The Birmingham, Alabama Fire Department bought a Seagrave combination truck in 1909. Horse fanciers set up a race between the new truck and horse-drawn equipment, but the race was inconclusive. A few days later, Birmingham-Southern College caught fire. While the horse-drawn equipment bogged down on a steep hill, the truck sped by and put out the fire.

A Caucus in Grand Rapids

The US Fire Chiefs' Convention in Grand Rapids, Michigan, was a turning point for the US. There, 550 fire chiefs from across the nation saw the demonstration of a new 650 gpm (2460 lpm) motorized pumper that cost $6500 from the factory.

They also heard the testimony of Captain John Glanville of St Louis. He said his motorized equipment cost just $481 per year to maintain versus $816 per year for his horses. The new trucks hurried to fires 50 percent faster and covered three times more of the city than the horse-drawn equipment.

Above and right: **This 1948 American LaFrance aerial ladder is maintained by the Goose Flats Fire Company of Tombstone, Arizona. Its 100-foot (30-meter) ladder is put to use almost daily in a variety of tasks.**

NO2 M. F. D.

Left: **This is an experimental, turbo-jet, gas turbine-powered American LaFrance pumper supplied to the San Francisco Fire Department in 1960. It proved to be unsatisfactory, with one objection being that heat from the overhead exhaust could ignite overhead awnings.**

Right: **At a muster one has the opportunity to see a variety of fire engines. In the foreground is a 1940s Mack, and in the background sits a 1941 Chevrolet operated by the San Francisco Fire Department.**

Below: **A 1950 FWD pumper with chromed grille.**

By the 1920s, manufacturers were swamped with orders for motorized equipment. The Savannah, Georgia Fire Department became the first fully motorized department in the US, buying seven pumpers, one chemical engine and four combination chemical/hose trucks from American LaFrance in 1911.

Though motorized equipment was becoming the rage, many municipalities held on to their horse-drawn apparatus, because in some cases there was simply too much investment in it, and in others—especially some locales in the peat- and coal-rich British Isles—it simply worked out better.

New York kept its horses until 1922 and Philadelphia kept its famous white stallions until New Year's Eve, 1929.

A Changing Technology

As the century progressed, so did the types of fires that had to be fought. Taller buildings required higher aerial ladders. Late twentieth-century ladder trucks are capable of mounting ladders that are extendable to 150 feet (45.6 meters) or more, often with a high-capacity nozzle built into its top section; some trucks carry 200 feet (61 meters) or more of ground ladders; hose trucks carry 1000 feet (300 meters) or more of hose.

Water towers were one way to fight upper-story fires: the name is self-descriptive—simply a tall, thin tower, bearing a hose and nozzle that could be aimed into an opening or window to access a fire.

BENSON
F.D.
DODGE
Job-Rated

In 1930, for instance, American LaFrance built a 65-foot (20-meter) tower with a capacity of 8500 gpm (32,175 lpm) for the fire department of New York. Pirsch topped that with an aluminum tower 100 feet (30 meters) high for the Melrose, Massachusetts Fire Department in 1925. American LaFrance then built one even higher in 1941, with a 125-foot (38-meter) tower for the Boston Fire Department.

Modern warfare expanded the scope of fire fighting. It started with the Zeppelin raids of World War I—the largest of which, on 8 September 1915, caused 29 fires and £530,000 in damage. World War II saw merciless bombing of Allied civilian targets by the Nazis, Japanese and Italians.

Brutal attacks in Asia and the Philippines were accompanied by artillery assaults on wholly unprepared populations; the British Isles were hammered continuously by the Nazis, who sought to break the people's spirit by relentless bombing.

World War II was a grim redefinition of the term 'urban fire.' Nazi Germany's London blitz included the first large-scale use of incendiary bombs on a modern city.

A few of the darker moments of the blitz are recounted here. Coventry, England, was the first modern example of the destructive capabilities of incendiary bombs—particularly, the raid of 14 November 1940, which reduced the city center to ashes. From 10 September 1940 through the year's end, German bombing raids

Left above: **This 1946 Ford-Howe pumper is equipped with a Barton-American front-mounted pump and a 400-gallon (1514-liter) water tank.**

Left below: **A mid-1950s Dodge-based fire engine.**

Below: **This open-cab Van Pelt pumper, built on a White chassis, looks as good today as the day it was delivered.**

ASTORIA FIRE DEPT.
ASTORIA
FIRE DEPT
SINCE 1870
GMC
FIRE DEPT.

Left above: **The Astoria, Oregon Fire Department's American LaFrance aerial.**

Above: **This early-1970s International pumper was outfitted by Ward-LaFrance.**

Left: **This 1968 Sutphen aerial tower served the Lockland, Ohio Fire Department. It was one of the first aerial platforms built by the company.**

on London destroyed the city's dock areas and caused 1500 separate fires. On 16 April 1941, 457 tons of incendiaries and high explosives were dropped over London, and on 10 May the House of Commons was destroyed.

Then came a turnabout, and the development of weapons spurred by the desperate, stubborn ferocity of the Nazis and their cohorts wrought utter devastation: the Allies took the war to them, with the intent to render Nazi Germany incapable of carrying on war. In addition to industrial Dresden, with its horrific firestorm, was Hamburg, which, from 24 July through 30 July 1943, suffered the destruction of 300,000 buildings and the loss of 60,000 to 100,000 citizens.

Terrible destruction and firestorms in Hiroshima and Nagasaki followed the nuclear bombings by the Allies.

What sort of equipment could cope with such raging fires? In most cases, the question was an afterthought, but the answer came as part of the entire parcel of the war—faster, bigger, more powerful.

FIRE ENGINES TODAY

The basic fire truck for any community is the pumper (or water tender), a truck that carries a water tank, ladders, hose, a pump and portable tools. Pumpers are usually designated 'first due, second due' and so on, in the case of big fires that require more than one pumper.

Upon arriving at a fire, one of the fire fighters hooks the truck up to a hydrant. (In extreme emergencies, they will use up the booster tank water first, and then go to a hydrant.) The chauffeur (as fire fighters like to call their drivers) then mans the pump control panel behind the cab to monitor water pressure through the hoses being used.

Booster tanks generally hold 500 gallons (1893 liters). An average medium-size pumper can put out 1500 gpm (5678 lpm). An average number of two hoses with 2.5-inch (6 cm) diameter have the capacity to spray 500 gpm (1893 lpm) over 1000 feet (305 meters).

Rural areas lacking fire hydrants or other readily available water usually have a tanker truck, which is equipped with a water tank averaging 1000 to 1500 gallons (3785 to 5678 liters) of water, a pump, hose and fire-fighting tools. Trucks normally used for brush fires are equipped with numerous small fire extinguishers and quantities of small-diameter hose.

Today's hook-and-ladder trucks carry ladders up to 135 feet (41 meters) in length which are mounted on a turntable, and stabilized by out-

Right: **Firefighters draw from their water tender while a Leyland DAF tanker truck stands at the ready.**

Below: **This Pierce Arrow rescue pumper is part of the Newtown, Pennsylvania fire and rescue fleet.**

Left: **An American LaFrance tractor-drawn, 100-foot (30-meter) aerial truck.**

Below: **The ladder on Nissan's 'Big Thumb' ladder truck can reach a height of 105 feet (32 meters).**

Below right: **This turntable ladder truck from Iveco Magirus is equipped with a rescue cage that can accommodate four people.**

rigger jacks on either side of the truck. Ladder trucks also carry hand tools.

The fire fighter at the top of the ladder directs water through the ladder's water nozzle and communicates via radio with the driver at the control panel. The rest of the crew head into the burning building with their tools, smashing doors, windows and other barriers to ventilation. This allows the smoke and poisonous gases to escape.

Reducing smoke aids in the search for people who are trapped, and reduces the danger of flareup or backdraft—which occurs when oxygen is freshly introduced to a smoldering ember.

Many hook-and-ladder trucks have steerable rear wheels, the evolution of which is explained earlier in the text. They are especially useful in cities with narrow streets and tight corners, but have been associated with traffic accidents, which is why some locales have phased them out in favor of hydraulic platform trucks. American LaFrance is one of the few companies that still make the hook-and-ladder.

Ladder tenders are a cost-effective alternative to the larger ladder trucks, as they do not have the expensive turntable arrangement, and yet provide a good array of fire-fighting tools for smaller communities.

The Snorkel aerial platform truck was invented in 1958 by Chicago Fire Commissioner Robert Quinn, who was inspired by watching workers maneuvering an elevated basket, otherwise known as a 'cherry picker.' Essentially a combination of a hydraulically-elevated basket and a length of hose, the aerial platform truck allows for stability and maneuverability that the ladder truck does not.

Aerial platform trucks also provide a more stable rescue platform than ladder trucks, because there's no water-soaked ladder the fire fighters could slip on. The baskets typically can

DAF
2500
INTERCOOLING
Angloco
G504 XFH

These pages: **This Angloco aerial is equipped with a Bronto Skylift 28-2T1 and is built on a Leyland DAF 2500 chassis.**

saval·kronenburg bv
MAN
1/24/2
1/24/2
FS·ED 820
M

LEEDS BRADFORD AIRPORT
4
4
Angloco

Left: **This heavy duty airport crash tender rides on a special chassis developed by MAN and the Federal German Army to operate in extreme off-road conditions.**

Below left: **This Angloco 8000 medium-sized air crash tender serves the Leeds Bradford Airport. The top-mounted turret can project a jet of foam up to 200 feet (61 meters).**

Below: **Iveco Magirus produces several rapid intervention vehicles. This model is capable of pumping 1056 gpm (4000 lpm) of either foam or water through its top-mounted cannon.**

support 900 pounds (408 kg) of weight. Aerial platform trucks can throw 1000 gpm (3785 lpm) through one or more turret nozzles.

Rehabilitation units are vehicles that offer fire fighters a misting system, providing relief from the intense heat of a fire, and giving them a chance to recoup after using up the compressed air in their fire-fighting apparatus.

Aircraft fire-fighting vehicles are usually found at airports and landing strips. They typically are equipped with foam and other specialized fire-fighting agents to handle aircraft fuel fires. Customarily, these vehicles are of two types—Rapid Intervention Vehicles, which weigh about 8000 pounds (3630 kg) and arrive on the scene quickly, holding off the fire with their five-minute allotments of foam; and the heavier, armored-car-like backup vehicles weighing 30 tons (27,000 kg) and sporting a turret gun that shoots copious plumes of foam.

Rescue trucks carry equipment for just about any emergency: air masks, resuscitators, chain saws, rope, chair stretchers, Hurst 'Jaws of Life' tools, blankets, first aid equipment, walkie-

talkies and more. Rescue squads have assumed a growing importance in today's complex world.

Mention of the equipment also requires comment on the agents used to fight fires. Water—modern technology has made it possible to add 'wetting agents' to it, enhancing its abilities to soak in; agents that produce a foam, which helps smother fire; or antifreeze for cold climates. Water is most commonly used on wood, paper and other substances that are comparatively easy to put out. Carbon dioxide is used wherever water can't be used to smother the flames. In common parlance, these are 'Class A' fires.

Class B fires involve flammable liquids which require a dry chemical, as do class C or electrical fires. Class D fires involve highly reactive metals like magnesium and phosphorus. Such fires are usually fought with dry powder.

Halogenated hydrocarbons, called halons, take the form of liquefied gas at room temperature, inhibiting chain-reaction combustion. Steam is usually used to control fire in confined areas, and inert gas is used to extinguish gas, dust and vapor fires.

Clockwise from right: **An International 4900 rescue vehicle; an Iveco Magirus emergency tender; a Leyland DAF 45 Series fire and rescue vehicle.**

GLEN ELLYN
VOLUNTEER FIRE CO.
GLEN ELLYN
GLEN ELLYN
SQUAD 38
38
Turbo
IVECO
MAGIRUS
OG
TZ 857
A
A

Carmichael

Carmichael Fire Ltd, established in 1849, is one of the oldest fire-equipment manufacturers in the world. Ability to build to exact customer specifications has deemed Carmichael one of the largest and most varied arrays of light vehicle offerings.

The firm builds its vehicles on a number of chassis, including Ford, Volvo, Scania, Dodge and Mercedes-Benz, for an extensive range of multi-purpose and light fire-fighting vehicles.

Carmichael also has a product support program, including training courses, preventative maintenance, emergency repairs and a full stock of spare parts for all vehicles manufactured by the company.

Carmichael's medium-capacity airport crash-support foam vehicles are often used on military airfields. The UK Ministry of Defense relies on Carmichael emergency vehicles such as the First Strike foam truck, and the Royal Air Force and Navy rely on its aircraft-support foam Crash Tender.

Civilian airfields also have recourse to specialty crash-support vehicles from Carmichael. Among its offerings are aerial trucks, ladder trucks and various tenders, plus multimedia water, foam, dry powder and carbon dioxide trucks.

Other offerings include the Redwing Light Pumping Apparatus, built on a Land Rover 4x4 chassis, and a high-speed First Strike foam truck on a converted Land Rover 6x4 chassis.

American LaFrance

Now based in Bluefield, Virginia, American LaFrance is part of the fire protection and safety conglomerate Figgie International, whose constituents also include Automatic Sprinkler, a specialist in building fire protection systems; Snorkel, maker of the Snorkel articulated basket; and Badger-Powhatan, a fire extinguisher manufacturer.

American LaFrance can trace its roots back to 1832, when it was then known as Rogers Patent Fire Engine Company. Among its notable achievements over the years are the first successful extension ladder truck (discussed previously in this text), invented by Daniel Hayes of the San Francisco Fire Department in 1868; the development of the Macomber Chemicalizer, the first pumper to combine water with other fire-fighting ingredients; and the company receives credit for building the first motorized hose and chemical truck in 1903.

The company built its last steamer in 1914, as it had already begun producing motorized trucks. From 1910 to 1926, American LaFrance produced over 4000 pumpers, plus ladder trucks, water towers and other apparatus, becoming the premier producer of fire-fighting equipment for decades to come.

Right: **This Carmichael multi-media extinguishing vehicle carries water, foam, dry powder and carbon dioxide and is based on a Mercedes-Benz 1936 4x4 chassis.**

Below: **These Carmichael fire trucks belong to the Welsh town of Llandundo. In the foreground is a turntable ladder built on a Dodge chassis, the other two trucks are water tenders, built on Dodge and Ford chassis.**

Clockwise from above: **This snub-nosed pumper serves the Fairfield, California Fire Department; one of several American LaFrance pumpers used by the San Francisco Fire Department; an American LaFrance aerial owned by the Astoria, Oregon Fire Department.**

Among its lasting contributions to the industry were the introduction of the first four-wheel fire engine brakes in 1929. They are also responsible for the introduction in 1939 of the first cab-forward fire engine, which would become an industry standard as the years wore on. American LaFrance, post-World War II, began making airport crash support vehicles, rescue trucks and airport structural fire support vehicles, as well as its other extensive lines of fire apparatus.

Indeed, American LaFrance is one name that comes to mind when the words 'fire truck' are said. Currently, the Century 2000 Series of vehicles is being produced.

Generally, 2000 Series trucks are ruggedly equipped for the rigors of their work: the entire cab, body and pump enclosure of each is low-carbon stainless steel, and the entire vehicle features modular construction, which means less down time for repairs. Cabs are available in four-door or two-door styles.

User-friendly control panels complement the 2000 Series setup, including a lifetime-guaranteed 500-gallon (1893-liter) fiberglass tank, and a dependable pump output of 1500 gpm (5678 lpm). All this is powered by a turbocharged 350 hp Detroit Diesel V-6, coupled to a an electrically-controlled Allison automatic gearbox and a rugged Rockwell R70 axle.

The Pacemaker series of vehicles encompasses American LaFrance's well-known cab and chassis stylés, the company's integral, all-stainless steel pumping modules, and stainless steel bodies.

The Water Chief and the Ladder Chief designations encompass American LaFrance's aerial trucks. These designations are available either as ladder trucks or as Snorkel trucks.

SAN FRANCISCO
ENGINE CO.

FAIRFIELD
AMERICAN LAFRANCE

SOUTH HACKENSACK
FIRE DEPT.
SOUTH HACKENSACK
FIRE DEPT.

These pages: **Engine No 3 of the South Hackensack Fire Department is a custom-built American LaFrance pumper.**

All models under these designations have a choice of chassis: Century 2000 or Pacemaker. The two aerial variations are 75 or 100 foot (23 or 30 meters).

The 75-foot (23-meter) Water Chief is usually a combination pumper and ladder truck, with up to 500 gallons (1893 liters) of water or a full complement of 163 feet (50 meters) of ground ladders, plus hand equipment.

The 100-foot (30-meter) Ladder Chief is available as a rear-mount aerial, a mid-mount or a tractor-drawn 'classic' variety. The rear-mount is the most popular at present, with the tractor-drawn aerials fading from the scene.

Snorkel variations on the Water Chief and Ladder Chief use a basket which supports up to 900 pounds (408 kg) and can extend to 85 feet (26 meters). Generally combined with the Tele-squirt Nozzle, the Snorkel tower can drench hard-to-reach flames with water. Water pressure is controlled from the base of the platform.

Perhaps someday the American LaFrance fire engines of today will achieve the same eminent collectors' status that their forebearers currently have. Since 1832, American LaFrance and its predecessor have been paragons of excellence in fire-fighting apparatus.

Seagrave

Seagrave has its origins in 1881, when laddermaker Frederick Seagrave was asked by a local Detroit fire department to make a vehicle to carry ladders to fires.

Seagrave turned out a handy two-wheel cart. On his own initiative, Seagrave then built a full-scale hook-and-ladder rig that turned out to be enormously successful.

Success brought a new venue and new ventures. After a move to Columbus, Ohio, Seagrave brought out an improvement on the

Left: **This American LaFrance pumper sits on a Pemfab Royale S-942 S/G Series cab and chassis. Pemfab has been supplying truck chassis to a great many fire apparatus manufacturers for over 20 years.**

Above: **A Seagrave tractor-drawn aerial with 100-foot (30-meter) ladder. The San Francisco Fire Department owns several tillered aerials.**

Right: **A rear view of the Seagrave tiller seat. The tillerman controls the rear axle to help the long truck navigate narrow city streets.**

Metz
DLK 23-12
1422
Freiwillige Feuerwehr
Stadt Sindelfingen
Metz
BB 242

Below: **the Mercedes-Benz DLK 23-12 turntable ladder is equipped with a hydraulically controlled ladder made by Metz Feuerwehrgeräte.**

Hayes aerial ladder mechanism in 1901. Dubbed a 'spring hoist,' Seagrave's invention revolutionized the aerial ladder industry.

Again, in 1912, Seagrave came out with a centrifugal pump that outpaced existing rotary pumps, setting another industry precedent. This was followed by an automatic regulator, which replaced the old, bulbous surge chamber for smoothing the water flow from pumps through hose; this in turn was followed by auxiliary coolers to keep the pumps running cool even after hours of work.

In 1935, Seagrave pioneered the first fully-hydraulic aerial ladder; and in 1939, a groundbreaking design allowed the crew to ride inside the engine cab, safe from the elements, to arrive fresh on the scene of any fire.

In addition to building safe equipment, Seagrave trucks are among the most aesthetically pleasing rigs in current service with fire departments.

Mercedes-Benz

In addition to having its heritage among the earliest internal-combustion auto designs, and being in the forefront of the luxury car industry today, this legendary German company is one of the most respected truck manufacturers in the late-twentieth-century world.

Supplying truck chassis for a wide variety of uses, including fire-fighting vehicle bases, Mercedes-Benz also puts out a full fire-fighting line of its own, including tender-pumpers, aerial ladder trucks, hydraulic platform trucks, tool and gear carriers, hose carriers, crew and fire chief transports, rescue vehicles and ambulances.

The backbone of the line is the Mercedes-Benz tender, which is available in various configurations. Tenders with water tanks, capable of pumping water and foam, or a combination of the two, are based on a wide variety of Mercedes-Benz commercial vehicles, fitted with a single- or multi-stage centrifugal pump. Capacities and outputs range from 264 to 2642 gallons (1000 to 10,000 liters) and 264 to 2114 gpm (1000 to 8000 lpm).

These include The LF 16 Fire Team Tender, built on a Mercedes 1222 AF/36 chassis and powered by a Mercedes OM 421 six-cylinder diesel. It delivers 634 gpm (2400 lpm) through its single-stage centrifugal pump.

The LF 16 water tank holds 317 gallons (1200 liters) and is fiberglass-reinforced plastic. It can carry up to nine fire fighters and is equipped with a rapid-intervention system, including a stable hose 100 feet (30 meters) in length.

The SPM/745 Pump Water Tender carries a crew of six, has an 1190-gallon (4500-liter) tank and delivers 600 gpm (2270 lpm). This model was specially built for operation in Tanzania. It features two rapid intervention sets, each with a 180-foot (55-meter) hose.

The GTLF 6 Large Capacity Pump Water Tender is powered by a Mercedes OM 422 eight-cylinder diesel, and delivers 740 gpm (2800 lpm) by means of its single-stage centrifugal pump. It can carry 1453 gallons (5500 liters) of water and 132 gallons (500 liters) of foam. A remote control water/foam monitor (also called a 'cannon') delivers 634 gpm (2400 lpm) of water with a throw of 180 feet (55 meters), or 634 gpm (2400 lpm) of foam with a throw of 157 feet (48 meters).

The FLF 11000/1350 Airport Crash Tender is a behemoth weighing 70,400 pounds (32,000 kg) and can hold 2906 gallons (11,000 liters) of water and 357 gallons (1350 liters) of foam. Its two-stage centrifugal pump delivers 1321 gpm (5000 lpm) of water through a roof-top monitor, with a throw of 246 feet (75 meters), and 1321 gpm (5000 lpm) of foam with a throw of 213 feet (65 meters). The pump is driven by a Mercedes OM 422 engine, while the truck is powered by a Mercedes 10-cylinder diesel.

Left: **This water tender is based on a Mercedes-Benz 1120 chassis and has a body by Albert Zeigler GmbH.**

Right: **This 6x4-wheel drive Mercedes-Benz universal fire tender has water, foam and powder capabilities. The body and equipment are by Zeigler.**

Below: **The hydraulic supports extending from the side of this Mercedes-Benz turntable ladder are part of the automatic 'Metz balance' protections. Apparently, the aerial is a useful tool in monorail maintenance as well as fire fighting.**

Tenders without tanks, capable of pumping from on-site sources, are generally equipped with a one- or two-stage centrifugal pump, and can carry, in the larger U 1300 L/37 model, 11 persons. A smaller tender, the 310/33, is also available. These vehicles carry protective clothing, small fire extinguishers, hoses, valves, ladders, safety blankets, rescue lines, resuscitation equipment, telecommunication equipment, hand tools and more.

Mercedes-Benz Foam Tenders are manned by a driver and two crew members. They are generally equipped with a high-capacity centrifugal pump to draw water from on-site sources, and are built on two- or three-axle heavy truck platforms to bear the weight of the 1057 to 2642 gallons (4000 to 10,000 liters) of foam. They also carry assorted protective clothing, fire-fighting tools, rapid-intervention equipment and valves.

Dry Agent Tenders dispense dry powder, carbon dioxide or halon to fight liquid and gas fires at such sites as refineries, chemical factories, airports and the like. Their powder storage tanks range from 1100 to 13,200 pounds (500 to 6000 kg) capacity. These vehicles are built on medium to heavy truck chassis, and the heavier versions (2200 pounds [1000 kg] capacity and up) discharge their agent through a monitor at rates up to 110 pounds per second (50 kg/sec), with throws of 148 to 246 feet (45 to 75 meters). Usually, there are auxiliary dry powder guns, complete with hoses, in addition to the manually-controlled monitor. The propellent is usually nitrogen or dry compressed air.

Mercedes-Benz Universal Tenders, which carry water, foam and dry agents, are equipped with single- or multi-stage centrifugal pumps with capacities of 264 to 2114 gpm (1000 to 8000 lpm), plus a water/foam monitor; a dry powder delivery system (including a monitor for vehicle capacities of 2200 pounds [1000 kg] and more); rapid intervention equipment; and assorted fire-fighting necessities. A middle-

Left, top to bottom: **A Mercedes-Benz airport crash tender; a Mercedes-Benz articulated and telescopic aerial with platform; a team of Mercedes-Benz medium class water tenders, the center truck has a front-mounted centrifugal pump.**

Right above: **This water tender is based on a Mercedes-Benz 1300 L, 4x4-wheel drive chassis.**

Right: **This Mercedes-Benz tender features bodywork designed by Gebr Bachert GmbH.**

Metz
UNIMOG
1300 L
KA-M 1842
1120
KA-2031

range Universal Tender may have a throw of 190 feet (58 meters) for water, and 174 feet (53 meters) for foam.

Mercedes-Benz aerials are available with water monitors. The DLK 44 K/F aerial ladder truck, for instance, has hydraulic ladders with a platform capable of reaching 144 feet (44 meters). The platform can hold 144 pounds (180 kg), and is equipped with a water monitor delivering 580 gpm (2200 lpm).

The SS 300 Snorkel has rate of flow controls in its basket, can rotate 360 degrees, reach 100 feet (30 meters), hold 803 pounds (365 kg), and deliver 1190 gpm (4500 lpm). Also available is a 238-gallon (900-liter) foam tank.

Fulton & Wylie

Founded in 1959, Fulton & Wylie Ltd moved from repairing commercial vehicle coachwork to building fire-fighting vehicles in 1968. Now the largest fire engine manufacturer in Scotland, the company uses a number of chassis makes—including Ford, Mercedes-Benz, Leyland DAF and Volvo—on which to base its vehicles.

For example, the Fulton & Wylie Leyland DAF Water Tender features a tilt cab large enough to allow fire fighters room to don their fire-fighting apparatus.

A 172 bhp Leyland turbocharged 420 engine can be had with an automatic or manual gearbox. The rear body usually features seven roller/shutter doors that allow for ample equipment storage, and can be fitted with a 264- or 475-gallon (1000- or 1800-liter) water tank and a Godiva pump.

Also available is a Volvo FL6-based Water Tender, powered by a Volvo TD 61F engine linked to an Allison gearbox for dependable performance.

The Type B Fire Wizard is based on the Mercedes 1120. The rear bodywork features the Fulton & Wylie pannier side that conforms to the contours of the cab. Monocoque construction helps reduce the noise level in the cab to extremely low levels, allowing for better prepa-

Left: **Fulton & Wylie Ltd of Irvine, Scotland puts this Leyland Freighter through a 'tilt test.'**

Right above: **A Fulton & Wylie 'Fire Warrior' water tender.**

Right: **From the front windscreen to the rear, this Mercedes-Benz 1120-based water tender was constructed entirely by Fulton & Wylie. It is owned by the Highland and Islands Fire Brigade.**

BEDFORD
1120
GAIRLOCH
GAIRLOCH SCHOOL
H536 AJS

ration en route to fires for the driver and crew. A 475-gallon (1800-liter) water tank, a Godiva GMA 2700 pump, Akron Marauder fog guns and two Collins Youldon hose reels complete the arrangement.

Oshkosh

One of the leading suppliers of airport crash and fire-fighting vehicles, Oshkosh supplies airports across the US and abroad.

Vehicles used in emergencies such as airport fuel spills or plane crashes must be capable of rapid deployment as well as powerful fire-suppression measures. This is a demanding set of criteria that also require carrying huge quantities of water and foam concentrate, as well as sophisticated on-board systems to mix the foam and deploy it with great vigor.

Named for the company's manufacturing home, Oshkosh, Wisconsin, Oshkosh once concentrated on all-wheel-drive vehicles for construction, highway building and snow removal. After World War II the advent of mass air travel opened a need to the company's guiding members, and today Oshkosh is at the top of the industry.

Oshkosh airport fire and emergency support vehicles include Rapid Intervention rigs that race to an accident scene for a preemptive strike. The company also has a line of unique low-level vehicles that are six feet (two meters) from roadbed to roof. They're meant for the low clearances common at airports.

Oshkosh also makes huge foam carriers that finish off fires with massive infusions of foam. These vehicles also have highly adequate off-road capabilities, to deal with rough terrain found at the perimeters of most airports.

Left: **This giant Oshkosh airport crash vehicle serves the US Air Force.**

Right: **A dependable and rugged vehicle, this Oshkosh crash tender is able to shoot a combination of extinguishing agents, thereby quickly controlling an airport emergency.**

Below: **Electronic controls make this powerful Oshkosh truck easy enough for one person to handle.**

Left below: **An Amertek CF4000L in use by the US Navy, part of the company's Aircraft Rescue and Fire Fighting team.**

Right below: **A field test of the Amertek CFRV-1, shown here fording a stream.**

Amertek

Ontario, Canada-based Amertek specializes in vehicles for the armed forces, specifically for fire-fighting at airports and military installations both at home bases and overseas.

The vehicles are code-named Aircraft Rescue and Fire Fighting vehicles, or ARFFs. Amertek's line includes the CFVR-1, a multirole fire fighter, for structural, fuel or brush fires. Featuring a roof-mounted turret monitor gun, the CFVR-1 is equipped with a 1000-gpm (3785-lpm) Godiva pump that can draw from an onboard 660-gallon (2500-liter) tank, from a relay pumper or from draft.

The vehicle can be driven onto and off of C130 and C141 airlift aircraft with ease, and is adept at offroading. Amertek vehicles figure prominently in US Armed Forces operations, including the 1991 Persian Gulf War.

A Rapid Intervention Vehicle manufactured by Amertek is the RIV-C1, used by airports across Canada. Powered by a Detroit Diesel Allison 6V92 TA engine, this large vehicle can go from 0 to 50 mph (80 kph) in 24.5 seconds, with a top speed of 73 mph (118 kph). Its roof-mounted turret gun dispenses foam at 330 gpm (1250 lpm) or dry chemicals at 15 pounds per second (6.8 kg/sec).

The US Navy uses the Amertek CF4000L, a 400-hp Detroit diesel-powered vehicle that carries 1000 gallons (3785 liters) of water and 130 gallons (492 liters) of foam, and is another Rapid Intervention Vehicle, with similar performance to the RIV-C1.

Angloco

Angloco was established as Angloco Coachbuilders Ltd in 1965, and moved to its present location in Batley, West Yorkshire, in 1972. The company's product line includes water tenders, emergency rescue vehicles, refinery tenders, all-terrain fire-fighting units, control units, aerials, breathing apparatus, hydraulic platforms, combined ladder platforms and airport crash and rescue equipment. Chassis used may be Ford, Mercedes-Benz, Scania, Volvo or others.

Municipal fire equipment from Angloco includes the Bronto Skylift 33-2T1 Combined Telescopic Ladder/Hydraulic Platform. A Volvo 318 bhp engine power drives the Volvo F1.10-based 33-2T1. Complete with stabilizing outriggers, these 8x4 trucks are favorites with the London Fire Brigade, which, at the time of this writing, has six of them. These are rescue units having a reach of 108 feet (33 meters), with a rescue cage of 880 pounds capacity (400 kg) at the end of an articulated boom. The cage can be swivelled 90 degrees, and has a remote-control monitor, stretcher support, and variable-voltage floodlight system.

Similar to this are the 28-2T1 outfits delivered to the Bedfordshire Fire and Rescue Service. These combine the Bronto rescue equipment with Scania P93.M1 chassis of a 6x4 configuration. Two crew members and the driver are accommodated in the cab.

The Oxfordshire Fire Service ordered a 28-2T1 with special features: an extended outreach, with a reduced cage capacity, allowing for a more versatile operation without re-positioning the truck; and a deck-mounted pump, which allows operation without a water tender, providing hydrants are available. Oxfordshire's 28-2T1 is based on a Volvo Fl.10 chassis.

Angloco is also the sole UK distributor of Metz turntable ladders. Metz Feuerwehrgerate, of Karlsruhe, Germany, makes 100-foot

Left: The Bronto Skylift 33-2T1 Combined Telescopic Ladder/Hydraulic Platform is manufactured by Tampere of Finland. A highly effective fire and rescue vehicle, Angloco is the sole UK distributor of the Bronto Skylift.

Above: This Bronto Skylift rides on a Volvo 6x4 chassis.

Right: The Oxfordshire Fire Service's Bronto Skylift is a 28-2T1 model. Additional features on this custom model include an extended outreach facility and a deck-mounted booster pump. The chassis is a Volvo FL.10.

Above: **Angloco is the sole UK distributor of Metz aerial ladders. This turntable ladder is built on a MAN 12-192 chassis.**

Right above: **This Angloco Emergency Response Vehicle is based on a Mercedes-Benz 1120F/36 chassis.**

Right: **Two Angloco water tenders on Leyland Comet chassis.**

(30-meter) telescopic ladders operable from the base controls or from the rescue cage. Metz ladders can be combined with stretcher support, a water monitor and floodlights.

Angloco specialist-support units include a Scania-based Foam/Carrier Tender, with 1700 gallons (6440 liters) of foam in a stainless steel tank. It's also equipped with a deck-mounted 868-gpm (3300-lpm) monitor, made to use on site-pressurized water supply in conjunction with foam from its own tank.

A 245 bhp engine drives the Scania 4x2 chassis, which is dressed in aluminum alloy bodywork, with ample storage lockers for hose and other fire-fighting equipment.

Angloco has also specially made such vehicles as the Emergency Response Vehicle, which incorporates features not usually found on 2x4-based fire equipment, also known as 'Type B' Water Tenders.

Roof-mounted rescue ladders and a rear locker arrangement are supplemented by a 1200-gpm (4540-lpm) pump, with an around-the-pump foam system. Two hose reels are also included, plus a backup 24-gallon (90-liter) foam-making apparatus, a 115-pound (50-kg) trolley-mounted hazardous-materials fire extinguisher and five breathing apparatus sets. The Emergency Response Vehicle can draw water either from the 480-gallon (1818-liter) onboard tank, or from any on-site water source.

Rapid Intervention Vehicles from Angloco include a unit built for the North Sea Gas Terminal of Total Oil Marine Plc at St Fergus. A 264-gallon (1000-liter) tank supplements a hydrant-feed system, via sidelines, hose reels and a dual-geared monitor with an output of 1150 gpm (4350 lpm) of foam/water solution. A 1321-gallon (5000-liter) foam tank is also supple-

FIRE
1120
BASF
FIRE AND EMERGENCY SERVICE
Angloco
Angloco
COMET
ST. LUCIA FIRE SERV
Angloco
COMET

Left: **An Angloco Rapid Intervention Vehicle built on a GMC chassis.**

Left below: **This Angloco Rapid Intervention Vehicle serves Total Oil's North Sea Gas Terminal at St Fergus.**

Right: **The Northumberland Fire and Rescue Service ordered this Angloco Type B water tender based on the Volvo F6.14 four-door cab extended crew chassis.**

Right below: **The Strathclyde Fire Brigade, Britain's second largest municipal brigade, operates this Angloco Type B water tender based on the Scania G82M four-door crew cab chassis.**

mented with a foam transfer system for feeding from outside sources.

Full operational capabilities are possible from the monitor mounted on the top rear of the body. The pump monitor can also control all vehicle systems. Two emergency, first-strike hose reels are also included. The vehicle is based on a Volvo F.16.17 chassis, with special hazardous-environment protection for the 207 bhp diesel engine.

Angloco also includes the Dodge 50 Series-based rescue/emergency tenders, with two side lockers and a full-width rear locker for ample carrying capacity.

For airport operations, Angloco's Type B Water Tender features a 2000-gpm (4540-lpm) rear-mounted pump, and a 480-gallon (1818-liter) tank. It features an 'around the pump' foam system allowing foam to be produced as needed. It also carries a full complement of ladders and equipment, including a portable pump, and two high-pressure hose reels, and can draw from its own tank, open water sources or hydrants.

The Angloco Water Tender is based on a Volvo FL6.14 chassis. Its cab tilts hydraulically, and has room for a driver and five crew members. Included in the cab are provisions for four BA sets and a telescopic light mast for flooding fire sites with light. A 207 bhp turbocharged, intercooled Volvo engine, linked to an Allison heavy duty transmission, provides motive power.

The Angloco 17000 Refinery Tender was built for Getty Oil of Kuwait. It carries 4491 gallons (17,000 liters) of foam, and is specifically designed for the inferno-like conditions of a major refinery fire. Based on a 38-ton variant of the all-

NORTHUMBERLAND
FIRE AND RESCUE SERVICE
VOLVO
FL6
Angloco
999
SCANIA
Angloco
82 M
HELLA

ROYAL AIR FORCE
FIRE AND RESCUE

Left: **Angloco built this airport crash fire/rescue tender for the Department of Civil Aviation in Malaysia. Based on the specialist Unipower 6x6 chassis with rear-mounted 540 bhp engine, this apparatus carries water and foam, and is capable of pumping 1200 gallons (4540 liters) of water per minute.**

***Left below:* The Hägglunds Bv 206 all-terrain carrier is ideal for transportation work that is inaccessible by normal wheeled vehicles. Angloco supplied this one to the Royal Air Force Fire and Rescue.**

***Below:* The Leeds Bradford Airport employs Angloco's 8000 Series air crash tenders.**

wheel-drive Mercedes-Benz 2636 chassis, it also has wide-section sand tires for the desert terrain.

The 17000 has hazardous area protection for its 355 bhp engine, air conditioning for its three-man cab, a public address system, a siren, warning beacons, a searchlight and other safety equipment.

Its PTO-driven pump can discharge up to 2400 gpm (9090 lpm) of water when drawing from hydrants, and can spew water or a combination of water and foam from six rear-mounted guns, as well as its roof monitor, thanks to the vehicle's 'balanced foam system.'

The Angloco Air Crash Tender 8000 is a Rapid Intervention Vehicle used on all terrain. Its water tank capacity of 1922 gallons (7274 liters), and foam tank capacity of 180 gallons (682 liters) ensure its potency when arriving on the scene, especially with a 1321-gpm (5000 lpm) Godiva centrifugal pump for the water and a 1009-gpm (4100-lpm) water/foam liquid delivery system.

The monitor can shoot a jet of foam 197 feet (60 meters), while at rest or in motion. This tender is built on a Boughton Theseus 4x4 chassis, powered by a General Motors turbocharged engine and an Allison gearbox.

Sutphen

Sutphen is a major custom fire apparatus manufacturer, with an emphasis on hand-built quality. For more than a century, this company has represented the highest quality in fire equipment. Based in Amlin, Ohio, Sutphen also has three other manufacturing locations. The main plant manufactures Sutphen 90- to 100-foot (27- to 30-meter) aerials and pumpers.

Sutphen builds its custom pumpers with a traditional two-door canopy cab; four-door safety cab with seating for six to 10; and a four-door tilt cab with seating for six to 10. Hale pumps and fiberglass water tanks complete these trucks. Body materials are optional: 12-gauge galvanized steel, extruded aluminum, or heavy duty bolted stainless steel.

The Hilliard, Ohio plant manufactures Sutphen Towers, especially the Mini-Tower, a 74-foot (22-meter) version of the 100-foot (30-meter) tower. Sutphen Towers are built of stainless steel, and stainless steel pumper orders are also processed at the plant.

The Sutphen 65-foot (20-meter) or 75-foot (23-meter) mini-towers are available with or without platforms. Equipped with a 1000- to 2000-gpm (3785- to 7570-lpm) Hale pump, 400- to 1000-gallon (1514- to 3785-liter) fiberglass tank and a full complement of ground ladders, these trucks are mounted on a 200-inch (5080-mm) chassis with a turning radius of just 378 inches (10 meters).

Midship-mounted design provides for reduced weight. Easy-access controls, a low profile and an electronically-controlled nozzle make for a setup time of just 20 seconds, with increased safety for fire fighters.

Springfield, Ohio hosts the Sutphen Chassis Division, where name-brand components are

Left: **A green Sutphen pumper delivered in 1992.**

Below: **This custom rescue vehicle is equipped to handle hazardous materials.**

Right above: **This four-door pumper built in 1989 has top-mounted pumps.**

Right: **This Sutphen custom tanker seats six to 10.**

STUART
STUART
FIRE RESCUE
ENGINE 11
FRANKLIN PARK
FRANKLIN PARK
V. F. C.
№-1
ENGINE 2

AMARILLO FIRE DEPT.
51
AMARILLO FIRE DEPT.
MT. LEBANON
25
MT. LEBANON

Above: **This Sutphen 75-foot (23-meter) mini-aerial ladder has the maneuverability of a pumper with the capability of a tower. The tilt cab provides easy access to the engine. The black vinyl pump panel is a standard feature on Sutphen mini-towers.**

Left above: **This 65-foot (20-meter) Sutphen aerial has twin platforms.**

Left: **Mt Lebanon, Pennsylvania, received this Sutphen aerial in 1992.**

assembled into Sutphen Custom chassis, from the frame rails to the final drive, every one rated as 'a deluxe custom chassis' by the company.

Monticello, New York is home for Sutphen East, the aerial ladder manufacturing and repairs/service/specialty apparatus wing of the company.

The efficiency of Sutphen aerials is such that five people can be rescued from a four-story building 25 seconds after the truck's arrival on the scene. Hale pumps, fiberglass water tanks and 206 cubic feet (six cubic meters) of storage, and centralized controls contribute to a 15-second setup time. An easy-to-maneuver four-section boom, and the fact that water can flow to the boom monitor even as it is being raised are contributing factors for a time efficient fire fighting operation.

All Sutphen trucks can be had with a choice of Detroit Diesel, Cummins Diesel or Caterpillar engines.

With the same name and same family ownership for over 100 years, Sutphen is the oldest fire engine manufacturer in continuous operation in the US. Even today, the next generation of the Sutphen family can be seen milling about the family factories after school.

Emergency One

A relatively new company, Ocala, Florida-based Emergency One was founded in 1974 as a new approach to fire truck construction—an almost wholly modular design, with emphasis on fuel efficiency. Warehousing truck chassis, pumps, ladder units and other equipment, Emergency One can custom-make a fire engine in as few as 60 days.

Custom and commercial pumpers, fast-response pumpers, platforms, aerials and telescoping booms, rescue vehicles and crash fire rescue vehicles are included in Emergency One's lineup.

Customers include the Whistler Fire Department, which oversees the number one ski resort in North America, at Whistler, British Columbia. Rescue considerations brought the department to order one of Emergency One's all-aluminum, 95-foot (29-meter) rescue platforms, with a narrow outrigger spread for tight situations. Fire departments all over the US and Canada have Emergency One equipment.

Above: **This Emergency One Hurricane pumper and matching 110-foot (34-meter) aerial are owned by the York County, Virginia Fire & Rescue Department.**

Left: **This Emergency One Hush Super Rescue unit equipped with an on-board computer in the command-style cab, operates in Bethlehem, Pennsylvania.**

Right: **Scarborough, Maine's Volunteer Fire Department depends on this Hurricane pumper. It is one of two in their service.**

L-1
YORK COUNTY
FIRE & RESCUE
COUNTY OF YORK
SCARBOROUGH
3
ENGINE CO.
PLEASANT HILL HOSE CO.

Right: **This 95-foot (29-meter) platform by Emergency One is built on a Hurricane chassis and has a Detroit 8V-92TA, 475 hp engine. The platform is able to hold 800 pounds (363 kg) of personnel with an equipment allowance of 250 pounds (113 kg). The Berwyn, Pennsylvania Fire Company is an all-Emergency-One fleet.**

BERWYN FIRE CO.
FIRE CO.
LADDER 2

Above: **This Emergency One Class A pumper is built on a Pemfab Imperial T-964 chassis. Pemfab Trucks is a major supplier of fire chassis, offering several cab forward and cab over designs.**

Left: **This Emergency One pumper features a tilt, 10-man, four-door fully enclosed sedan cab and is built upon a Pemfab T-944A chassis.**

Right: **This Emergency One Rescue vehicle is built upon an International 4900 Series chassis.**

The company's Hush XL pumpers have a rear-engine design to increase room and reduce noise in the cab: at 78 decibels, these cabs can be used as command centers en route to fires, and with so much room, can be used as rest stations by personnel at the scene. Seating for 12 is allowed.

Low cab steps make for more safety and convenience, and short, 160-inch (4 meter) wheelbases enable Hush XL pumpers to maneuver with ease.

Protector XL Series pumpers have 500- to 1000-gallon (1893 to 3780 liter) aluminum water tanks and single or two-stage midship pumps of 100 to 2000 gpm (3785- to 7570-lpm) capacity.

Heavy-duty Cyclone or Hurricane chassis have high-horsepower engines for more dependability and quicker response time, and a four-door, fully-enclosed cab with seating that accommodates eight.

Emergency One commercial pumpers range from 500 to 1500 gpm (1893 to 5678 lpm) capacity, with 500- to 1250-gallon (1893- to 4732-liter) aluminum water tanks.

A new product combines a 75-foot (23-meter) aerial with 1500-gallon (5678-liter) pumper equipment.

The company's 135-foot aerial tops a line of 75-, 80-, 96- and 100-foot (23-, 24- 295- and 30-meter) aerials; 95-foot (29-meter) platforms; and 50-foot (15-meter) telescoping booms. All of these are equippable with an optional pump and water system.

High-strength, non-corrosive aluminum, an automatic leveling system and a pedestal control panel characterize the 95-foot (29 meters) platform. The 50-foot (15 meters) telescoping boom has three sections, allowing for short wheelbases (153 to 200 inches/3886 to 5080 cm) with no boom overhang. A delivery system with 1000 gpm (3785 lpm) capacity is average.

Ford

As one of the world's leading truck manufacturers, Ford has also led the way among the big three automakers in the manufacture of fire trucks and chassis for fire trucks. From the days of the Model A, Fords have been in demand by fire departments around the world. The old Model T chassis were often used as a basis for American LaFrance equipment.

With the introduction of the Ford V-8 in the 1930s, Ford trucks have been in even greater demand. The mid-1950s F-750, equipped with an overhead-valve update of the old Ford V-8, was one of the most popular of all chassis for fire equipment.

The Ford F-Series and C-Series (especially the C-900 chassis) remained popular with fire engine manufacturers through the 1960s and 1970s. This tradition has continued into the 1990s, with Ford providing chassis for numerous apparatus builders, including Darley, Emergency One, Grumman and others.

Grumman

Grumman established a great reputation in the aircraft industry, most notably with its series of aircraft carrier-based fighter/dive-bomber planes in World War II. These prop-driven planes were known as 'carrier cats' because of Grumman's habit of attaching the suffix 'cat' to its products. The most famous of the World War II products was the 'Hellcat.'

Grumman bought out the legendary Howe Fire Apparatus Company in the late 1960s and used its aviation experience to design and build truck and bus bodies.

Predictably, even in the fire-fighting business, Grumman's 'cats' live on—from the Minicat to the Firecat to the Aerialcat.

Above: **A commercial pumper built on the popular Ford Cargo chassis.**

Right above: **Ford's F-Super Duty truck provides the basis for this light pumper. Bodywork is by Bridgewater Metal of Canada.**

Right: **This pumper is based on a Ford L-9000 chassis.**

VOLUNTEER
FIRE DEPT.
EEP BACK
150 m.
R20592
L9000

Grumman's tanker is called Tankercat, and is capable of supplying 1200 to 1500 gallons (4540 to 5680 liters) with a 250-gpm (950-lpm) pump. Its big brother is a super pumper-tanker, with galvanized tanks holding 2000 to 3000 gallons (7570 to 11,360 liters), and pumps capable of 2000 gpm (7570 lpm). The tankers have modular bodies, available in galvanized steel or aluminum.

The Grumman Aerialcat is one of the top-selling heavy-duty aerials in the world, and number one in the US. Its working height is 95 or 102 feet (29 or 31 meters), with a platform payload of 1000 pounds (453 kg) for the 95-foot version, and 800 pounds (362 kg) for the 102-foot version. A rear-mounted, 85- or 106-foot (26- or 32-meter) steel ladder is also offered, touted by the company as well suited to tight city conditions.

Grumman's stock pumper, the Wildcat, offers a variety of chassis, 750- or 100 gpm (2840- or 3780-lpm) Waterous pumps, and 750- or 1000-gallon (2840- or 3780-liters) booster tanks.

The Firecat and Tigercat are the center of the Grumman offerings, featuring modular design and a galvanized steel body for the former and an aluminum body for the latter. Both have optional chassis and drivetrain selections, and feature 750- to 1500-gpm (2840-to 5680-lpm) Waterous pumps with a 600- to 1000-gallon (2270- to 3780-liters) galvanized booster tank.

Left: **A Grumman Firecat pumper belonging to the Oakland, Maryland Volunteer Fire Department. It rests on a Pemfab Royal S-944A Series chassis.**

Above top: **Grumman's 95-foot (29-meter) Aerialcat.**

Above: **This Grumman Wildcat stock pumper is shown on a Ford F-800 chassis.**

The Attackcat is a quick-response, mid-size pumper that, while shorter than the average pumper by two feet (.75 meter), yet has a full-size hose bed and a water output of 500 to 1000 gpm (1890 to 3780 lpm), with a single or a two-stage Waterous pump, and a 350 or 500 gallon (1330 or 1890 liter) tank.

A quick-attack pumper, the Minicat, puts out 250 gpm (950 lpm) with its Waterous pump, and carries 250 gallons (950 liters) in a galvanized steel water tank.

The smallest Grumman 'cat' is the Skiddycat, available as a slide-on unit for pickup trucks or trailers.

Darley

Though it is a small company, WS Darley & Co, of Chicago, Illinois, has a worldwide reputation for delivering exactly what community fire departments have ordered across the US and in 62 other countries.

Darley is the only fire equipment company that builds both the fire engine body and the pump. The heart of every Darley pumper is the Darley pump, engineered and tested for the specific chassis selected.

Engine power is likewise carefully calibrated for maximum effect through careful research of gear ratios and combinations.

The Monarch Pumper Series of trucks represents one of two standard lines (the other being the Challenger line) that supplement the company's custom lines. Monarchs are available in 500 to 1250 gpm (1893 to 4732 lpm) capacities, with tilt-cab accessibility to engine parts.

The Challenger line includes a Darley 1000 gpm (3785 lpm) two-stage pump; pump relief valve with flush system; manual shutoff, self cleaning strainer and discharge check valve; dual pilot lights; and a 750-gallon (2840-liter) booster tank, warrantied for 10 years.

In all, Darley offers a range of equipment including pumpers; minipumpers; tankers; drop-in units for pickup trucks; trailers with pump, tank and controls; rescue vehicles; variable-mount pumps; portable pumps; and a choice of standard or deluxe chassis for pumpers.

Left: **A Darley rescue vehicle.**

Below: **This Darley pumper is built on a KME chassis.**

Right above: **This Darley pumper built on a Spartan chassis serves King County, Washington.**

Right below: **A Darley body and pump on a Ford F-800 chassis comprise this pumper.**

ENG.
Emergency / Dial
911
KING COUNTY
FIRE DIST. No. 2
COLERIDGE
ERECT
V.F.D.
101
DARLEY

Above: **Mack Trucks built this pumper for the Fayetteville, Pennsylvania Volunteer Fire Department. It is based on a Mack CH600 chassis.**

Left: **Six Mack pumpers for the Scarborough Fire Department with bodywork by Thibault.**

Right above: **This pumper is based on a Mack MC fire chassis. It features a Waterous 1500 gpm (5678 lpm) pump and bodywork by Pierce.**

Mack

Mack Trucks, of Allentown, Pennsylvania, has a reputation for toughness and reliability that few other manufacturers can rival. The company was the first to build a large-capacity triple combination pumper in 1935. This machine put out 2000 gpm (7570 lpm).

Its most astounding achievement came in the mid-1960s, however, when Mack custom-built a five-vehicle, 18-wheel Super Pumper System for the Fire Department of New York. The system was designed to douse large, rapidly burning buildings before a true conflagration could begin. It was delivered to the City of New York in October 1965 for $875,000, and soon proved its worth during a five-alarm fire at a Bronx candy factory. The Super Pumper's four and one-half-inch hoses surged with 8800 gpm (33,310 lpm) of water as Satellite Number Two's huge water cannon brought the inferno under control in 10 minutes.

The heart of the system was a 34-ton tractor-trailer powered by a 2400-hp Napier-Deltic diesel engine. A six-stage DeLaval, centrifugal pump (adaptable to either fresh or salt water), was made of stainless steel. Attached was the Super Tender, a tractor-trailer with 2000 feet (610 meters) of super-heavy-duty hose.

A giant water cannon behind the Super Tender's cab could shoot a thick stream of high-pressure water 600 feet (183 meters) high. Three satellite tenders, delivering 2000 gpm (7570 lpm) completed the team.

Mack is also credited with the first diesel fire engine, the first front and rear fire engine disc brakes and the first four-wheel-drive engine chassis.

Mack supplies its designs with its own power trains, but customers can special-order such engines as Caterpillar or Detroit diesel—the way to tell from the outside is, the signature bulldog ornament is gold if the truck is 'solid Mack,' and silver if another maker has supplied the motive power.

Mack's telescoping Aerialscopes are ideal for confined city streets, and are said to be more maneuverable than articulated booms.

Pierce

Pierce began making fire engine bodies in the late 1930s, and by 1979 had progressed to the production of all-aluminum cabs for its bodies. That same year, the company acquired the rights to the name 'Pierce Arrow,' once the designation of one of the most luxurious cars in the world. The Pierce Arrow name was applied to the sleek, new line of aluminum cabs.

In 1984, the Pierce Dash was introduced—a new cabover configuration with a four-seat capacity. A companion line, the D-8000, is, like the Dash, a pumper with a mid-mounted Waterous pump and a 5000- to 1000-gallon (1893- to 3785-liter) water tank.

The Pierce Arrow and the Pierce Suburban Series are popular still today, with a choice of several Detroit diesel engines of up to 500 hp.

The Javelin is Pierce's 'ultimate high-performance fire-fighting machine.' Equipped with front-wheel drive, a mid-ship mounted Waterous pump and 500-, 750- or 1000-gallon (1893-, 2840- or 3785-liter) tank, and having room for 11 fire fighters in its cab, the Javelin is a fine piece of equipment.

A 100-foot (30-meter) aerial platform and aerial ladders of 55, 75, and 105 feet (17, 23 and 32 meters), mountable on any of Pierce's custom chassis, combine with a wide range of rescue vehicles to complete the impressive Pierce lineup.

Left: **The Dash pumper by Pierce.**

Below: **A 65-foot (20-meter) Tele-Squirt aerial ladder and water tower, and a 100-foot (30-meter) aerial platform. Both are mounted on Pierce Arrow chassis.**

Right above: **A Pierce Javelin pumper.**

Right below: **This Pierce Suburban is part of their Responder Program.**

Overleaf: **Pierce's six-man split, tilt-cab Lance pumper.**

METRO-DADE
FIRE RESCUE
METRO-DADE
FIRE RESCUE
WEST
WILDWOOD
FIRE Co

ENGINE NO. 1
TOWN OF ODON
FIRE DEPT.

Simon

Simon Access, a division of the Simon Group, is the world's leading supplier of crash, fire and rescue vehicles and aerial platforms. Simon equipment is used by firefighters in over 70 countries around the world.

The Simon Pacer line is a series of first-strike vehicles built to be comparatively light and fast. Built on GMC all-wheel-drive chassis, Pacer cabs can be modified as necessary. The line features a Rapid Intervention Vehicle with a 317-gallon (1200-liter) water tank and a 40-gallon (150-liter) foam tank. In addition, there is a Pacer Powder truck, with a 2200 pound (1000 kg) capacity.

The Simon Defender Line can be based on any suitable 6x6 or 4x4 chassis, with or without extended cabs. The 4x4 version carries 1190 gallons (4500 liters) of water, and can deliver 793 gpm (3000 lpm); while the 6x6 version can carry 1850 gallons (7000 liters) of water and 222 gallons (840 liters) of foam, delivering 925 gpm (3500 lpm) via its Godiva pump.

The Defender RIV can attain 55 mph (80 kph) in 25 seconds or less, and can put out 635 gpm (2400 lpm), holding 634 gallons (2400 liters) of water and 143 gallons (540 liters) of foam.

Right: This Simon Protector C3 serves the Belfast City Airport of Northern Ireland.

Right below: Simon's Defender 4x4 is a crash, fire and rescue vehicle with water and foam capabilities.

Below: The Simon Pacer has an enclosed storage area and an elevating mast.

SIMON
ELFAST CITY
-AIRPORT-
PROTECTOR

The Airport
Isle of man
4

The Protector line includes the massive C3 and C4 vehicles, with 2642 to 3170 gallons (10,000 to 12,000 liters) of water and 317 to 383 gallons (1200 to 1450 liters) of foam for the C3 and, for the C4, 3170 to 3700 gallons (12,000 to 14,000 liters) and 383 to 450 gallons (1450 to 1700 liters), respectively.

The Protector C2 has a capacity of 1057 to 1585 gallons (4000 to 6000 liters) of water, and a foam capacity of 127 to 190 gallons (480 to 720 liters). Designed to be the follow-up to a Rapid Intervention Vehicle, the C2 manages a respectable 55 mph (80 kph) in 30 seconds.

The Protector RIV accelerates to 55 mph (80 kph) in 25 seconds or less, has a crew of three, and carries 1057 to 1585 gallons (4000 to 6000 liters) of water and 127 to 190 gallons (480 to 720 liters) of foam. This rapid intervention vehicle is designed to reach the scene of an emergency before a worst-case scenario breaks out.

Simon offers a wide range of aerial platforms including the 26 Series, the SCA Series and the ST Range. The 26 Series and SCA Series models are boomed, articulated, Snorkel platforms with varying operating heights. The 26 Series models are designed with three booms whereas the SCA Series models utilize only two.

The ST Range are telescopic Snorkels incorporating state-of-the-art technology which makes them the most maneuverable aerial platforms ever produced.

In 1984, Simon introduced the record-breaking SS600 Super Snorkel. This telescopic Snorkel was the first low-profile, high performance appliance with a 200-foot (62-meter) maximum working height.

Left: **Simon's breath-taking SS600 Super Snorkel can reach a record-breaking height of 200 feet (61 meters).**

Right: **The articulated SS300 hydraulic platform (right) works alongside a telescopic ST300 (left), showing the Simon's units' different boom configuration and jacking styles.**

Overleaf: **The Simon ST300 telescopic boom ladder is seen in its compact, travelling position.**

saxon
SWL 400KG

CENTRAL

Left: **This Kenworth pumper serves the City of San Francisco. The low, cab-forward design allows for easy access and improved visibility.**

Right: **This Kenworth T800 tanker is compact yet rugged.**

Below: **This gleaming Kenworth W900 tanker is part of the Whatcom County, Washington Fire District fleet.**

Kenworth

Kenworth is a major long-haul truck manufacturer whose products are popular around the globe. Kenworth started making fire apparatus in 1932, and in 1935 the company's first ladder truck was built.

In 1939, a cabover truck for the Los Angeles Fire Department was considered to be revolutionary for the fire vehicle industry.

In 1940, Kenworth built the world's first fully automatic aerial truck for the Spokane Fire Department.

A 1951 Kenworth pumper was one of the first designs to include a cab assembly fully fabricated by Kenworth, and a 1952 'Bullnose' model featured a squared-off version of the cabover versions of the 1930s.

In 1965, Kenworth introduced a tank truck with a front-mounted low-capacity pump for extra versatility. This model was in production until 1987.

In 1980, the company's low cab-forward design afforded personnel easier access and better visibility. The 1990s will no doubt see more innovative designs from Kenworth.

Peterbilt

Though instituted as a log-truck company, one of Peterbilt's first customer requests was for a fire engine—a pumper that was displayed at San Francisco's Golden Gate International Expo, and was later sold to the city of Centerville, California.

Through the 1950s the company supplied chassis to such fire apparatus builders as OH Hirst, Challenger Fire Equipment, PE Van Pelt and other firms, virtually all the orders going to locations in California. The firm's logging expertise also stands up well in equipment built to fight forest fires.

The Peterbilt Model 320 has become the model of choice for fire apparatus produced by the company. Its low cab-forward design is ideal for fully-equipped and heavily-clothed fire fighter access.

Another favorite is the big, conventional-cab Model 377, with state-of-the-art aerodynamic styling. Peterbilt, though well-respected, maintains only a small output of fire-fighting vehicles—perhaps 25 to 50 out of a yearly production run of 10,000 to 15,000 units overall.

Left: **Peterbilt's Model 377 is the muscle behind this pumper.**

Below: **Two pumpers on the Peterbilt Model 359 chassis.**

Above: **This Peterbilt cab-over pumper with bodywork by Van Pelt is easier to enter for fire fighters in full gear.**

Right: **Peterbilt's flagship chassis Model 359 is the basis of this gleaming pumper.**

INDEX

PICTURE CREDITS

AGS Picture Archives: 14, 15 (both), 16, 17 (top), 23, 28
American LaFrance Division, Figgie International: 70-71
Amertek: 84-85 (all)
Angloco Ltd: 8-9, 62 (bottom), 86, 87 (both), 88, 89 (both), 90 (both), 91 (both), 92 (both), 93
Bridgewater Metal Industries: 105 (top)
Carmichael Fire Ltd: 67 (top)
CIGNA Museum and Art Collection: 21, 27
John Creighton: 66-67 (bottom)
WS Darley & Co: 108 (both), 109 (both)
© RE DeJauregui: 30-31, 40 (both), 58 (top), 68 (top), 69 (top), 73 (both), 122 (top)
© Reg Dunham: 22, 24-25, 29, 36, 37, 41 (top), 46 (both), 52 (bottom), 53
Emergency One Inc: 11 (bottom), 98-99 (all), 100-101
Ford Motor Co: 2, 104, 105 (bottom), 107 (bottom)
Fulton & Wylie Ltd: 80, 81 (both)
Grumman Emergency Products: 107 (top)
Iveco Magirus AG: 10, 59 (right), 63 (bottom), 65 (bottom)
Kenworth Truck Co: 122 (bottom), 123
© James L Kidd: 42, 43, 44-45, 48, 49, 52 (top)
Leyland DAF: 57 (top), 60-61, 64
Mack Trucks Inc: 110 (both), 111
MAN Nutzfahrzeuge AG: 62-63 (top)
Mercedes-Benz AG: 74-75, 76, 77 (both), 78 (all), 79 (both)
Navistar International: 65 (top), 102 (bottom)
Nissan Diesel Motor Co Ltd: 4-5, 58 (bottom)
Oshkosh Truck Corporation: 83 (both)
Pemfab Trucks Division: 1, 72, 102 (top), 103, 106
Peterbilt Motors Co: 3-6, 124 (both), 125 (both)
Pierce Manufacturing Inc: 56-57 (bottom), 112 (both), 113 (both), 114-115
John R Schmidt via Sutphen Corp: 96 (bottom)
Simon-Dudley Ltd: 118, 119, 120-121
Simon Gloster Saro Ltd: 116, 117 (both)
© Leonard Smith: 13, 17 (bottom), 18-19, 26, 33 (top), 34, 35
Sutphen Corporation: 54 (bottom), 94 (both), 95 (both), 96 (top), 97
US Department of Defense: 11 (top right)
Volvo GM Heavy Truck Corporation: 11 (top left)
Ward-LaFrance: 55 (top)
Western Reserve Historical Society: 32-33 (bottom)
© Don Wood: 38 (both), 39, 41 (bottom), 47 (top), 50 (both), 51
© Bill Yenne: 54 (top), 69 (bottom), 82

***Below:* This Leyland DAF water tender serves the Lancashire County, England Fire Brigade.**